DEPARTMENT OF THE NAVY
HEADQUARTERS UNITED STATES MARINE CORPS
WASHINGTON, D.C. 20380-1775

11 October 2012

FOREWORD

Marine Corps Reference Publication 3-01A, *Rifle Marksmanship*, presents how the Service rifle/carbine is employed by the individual Marine, contains the skills and techniques required for individual rifle marksmanship, and provides the skills that are required for basic through advanced marksmanship.

This publication is intended to be used as a reference guide concerning marksmanship skills for unit commanders, trainers, and individual Marines. It presents marksmanship techniques organized by topic, and within each of those topics, techniques are broken down further as they apply to employment with the rifle combat optic (the primary sighting system) and with backup iron sights.

This publication supersedes Marine Corps Reference Publication 3-01A, *Rifle Marksmanship*, dated 29 March 2001.

Reviewed and approved this date.

BY DIRECTION OF THE COMMANDANT OF THE MARINE CORPS

RICHARD P. MILLS
Lieutenant General, U.S. Marine Corps
Deputy Commandant for Combat Development and Integration

Publication Control Number: 144 000091 00

THIS PAGE INTENTIONALLY LEFT BLANK

Rifle Marksmanship

Table of Contents

Chapter 1. Introduction to Rifle Marksmanship

Chapter 2. The Marine Corps Combat Marksmanship Program

Chapter 3. The Service Rifle

Chapter 4. Weapons Handling

Chapter 5. Fundamentals of Marksmanship

Chapter 6. Service Rifle Firing Positions

Chapter 7. Effects of Weather

Chapter 8. Zeroing the Service Rifle

Chapter 9. Offset Aiming for Windage and Elevation

Chapter 10. Engagement Techniques

Chapter 11. Target Detection and Range Estimation

Chapter 12. Multiple Target Engagement Techniques

Chapter 13. Moving Target Engagement Techniques

Chapter 14. Movement

Chapter 15. Low-Light Engagement Techniques

Appendices

Glossary

References and Related Publications

CHAPTER 1

Introduction to Rifle Marksmanship

All Marines share a common warfighting belief: Every Marine a rifleman. This simple credo reinforces the belief that all Marines are forged from a common experience, share a common set of values, and are trained as members of an expeditionary force in readiness. Therefore, there are no "rear area" Marines, and no one is very far from the fighting during expeditionary operations. As in the past, the Marine rifleman will be among the first to confront the enemy and the last to hang his weapon in the rack after the conflict is won.

Marine Corps forces are employed across the entire range of military operations. Those operations include war (characterized by large-scale, sustained combat operations) and other military operations that focus on deterring aggression, resolving conflict, promoting peace, and supporting civil authorities. These operations can occur before, during, and after combat operations.

Marines always need to be prepared to carry out their mission and adapt to changing levels of threat and unexpected situations. Whenever the situation warrants the application of deadly force, the Marine rifleman must be able to deliver well-aimed shots to eliminate the adversary. Marines must have the versatility, flexibility, and skills to deal with a situation at any level of intensity across the entire range of military operations (see fig. 1-1).

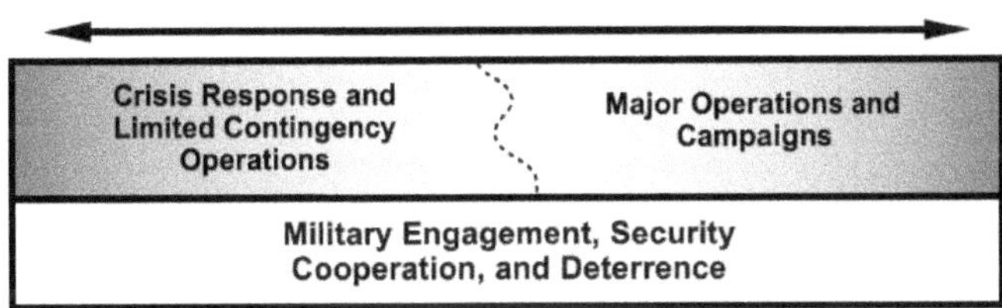

Figure 1-1. Range of Military Operations.

To be combat ready, the Marine must be skilled in the techniques and procedures of rifle marksmanship and take proper care of his Service rifle. To send Marines into harm's way without thorough training in the use of their individual weapons carries undue risks for every Marine in the unit.

THIS PAGE INTENTIONALLY LEFT BLANK

CHAPTER 2

The Marine Corps Combat Marksmanship Program

Combat-ready Marines must be skilled in tactics and proficient in the use of firearms. Well-trained Marines will have the confidence required to deliver accurate fire under the most adverse battle conditions.

The Service rifle is the primary tool that an individual Marine uses to obtain dominance over the enemy and neutralize the effects of enemy weapons. The Marine Corps Combat Rifle Program progresses the individual Marine from the fundamentals of marksmanship to advanced combat shooting by ensuring that Service standards are reviewed, practiced, and evaluated. The objective at the completion of training is that the Marine will have the ability to competently and effectively employ the Service rifle under varying conditions.

> ***Note:*** The Service rifle is defined as either the M16A4 rifle or M4 carbine equipped with a rifle combat optic (RCO). Hereafter, the term "Service rifle" will be used to refer to both the M16A4 with RCO and the M4 with RCO, as defined in Marine Corps Order (MCO) 3574.2_, *Corps Combat Marksmanship Programs*.

The Marine Corps Combat Marksmanship Program includes the Marine Corps Combat Rifle Program, the Combat Pistol Program, and the Marine Corps Competition in Arms Program.

Marine Corps Combat Rifle Program

The Marine Corps Combat Rifle Program uses a building block/training continuum approach toward developing Marines into proficient combat marksmen. To accomplish this, the Marine Corps Combat Rifle Program is broken down into different stages of training using the firing training tables found in MCO 3574.2_.

Marines will begin the program by learning the fundamentals of marksmanship (using the preparatory training in tables 1 and 1A) and proceed incrementally into new combat shooting skills (using tables 2 through 4). Each table of training must be mastered and completed in sequential order.

As the Marine Corps Combat Rifle Program progresses, formal training is included on modular attachments (e.g., optics, night aiming devices) that are associated with the Service rifle.

During preparatory training, weapons handling and zeroing skills are developed and mastered. Preparatory training prepares the Marine for the following tasks that are associated with a Service rifle:

- Performing weapons handling procedures.
- Maintaining the weapon.
- Performing corrective actions.
- Zeroing the weapon.

Fundamental Rifle Marksmanship Firing Tables

Tables 1 and 1A include the information that is necessary for the Marine to develop the fundamental knowledge, skills, and attitudes that are required to be safe and accurate when firing the Service rifle. The information covered in this stage of training forms the basis for all other training and includes performing the following tasks with a Service rifle:

- Zeroing a RCO to a Service rifle.
- Engaging targets from a prone position.
- Engaging targets from a sitting position.
- Engaging targets from a kneeling position.
- Engaging targets from a standing position.
- Engaging targets at the sustained rate of fire.
- Performing the fundamentals of marksmanship.

Basic Combat Rifle Marksmanship Firing Tables

Basic combat rifle marksmanship is the first step in transitioning the Marine from fundamental marksmanship to becoming a proficient combat marksman. When performing table 2 firing tables (i.e., training, preevaluation, and evaluation) the following tasks are trained with a Service rifle:

- Demonstrating weapons carries.
- Executing a tactical reload.
- Executing a speed reload.
- Executing controlled pairs.
- Executing failure to stop drills.
- Executing multiple target engagements.
- Engaging a moving target.

Intermediate Combat Rifle Marksmanship Firing Tables

Intermediate combat rifle marksmanship training (tables 3A through 3D) reinforces and improves basic combat shooting skills and introduces additional techniques and procedures. Upon completion of intermediate combat rifle marksmanship firing tables, the Marine—

- Will have demonstrated the required skills for successful completion of the tasks assigned to the Marine rifleman.
- Will be satisfactorily prepared for additional mission-specific rifle training as determined by the commander.
- Will be introduced to modular attachments, such as RCO and night aiming devices, if available to the unit during this stage of training.

When performing these tables, the following tasks are trained with a Service rifle:

- Zeroing a RCO to a Service rifle.
- Zeroing a target pointer illuminator or aiming light.
- Executing hammer pairs.
- Engaging targets using pivot techniques.
- Engaging targets while moving forward.
- Engaging targets at night.
- Engaging targets using a target pointer illuminator or aiming light.
- Engaging targets from an unknown distance.

Advanced Combat Rifle Marksmanship Firing Tables

The advanced combat rifle marksmanship training reinforces and improves combat shooting skills and introduces advanced techniques and procedures relevant to the infantry Marine. Upon completion of advanced combat rifle marksmanship firing tables (tables 4A and 4B), the Marine will—

- Have demonstrated the required skills for successful completion of the rifle tasks that are assigned to the infantry Marine.
- Be satisfactorily prepared for additional infantry-specific rifle training as determined by the commander.

> ***Note:*** Advanced techniques for modular attachments, such as RCO, night aiming, and night vision devices, are continued and improved during this stage of training while using lateral movement techniques.

Marine Corps Competition in Arms Program

The purpose of the Marine Corps Competition in Arms Program is to—

- Develop and maintain a population base of Marines with high skills in rifle and pistol marksmanship.
- Stimulate interest and desire on the part of the individual Marine for self-improvement of skills and confidence with both the Service rifle and the pistol.
- Gain and maintain the Marine Corps' ability to compete and win, as both teams and individuals, in interService and national matches and to provide competitors for United States teams in international matches.
- Establish a vehicle for the development and exchange of ideas resulting in improvements to equipment and shooting techniques.

CHAPTER 3

The Service Rifle

The Service Rifle

Description

The Service rifle is defined as either the M16A4 rifle or M4 carbine equipped with a RCO. The Service rifle is a lightweight, 5.56-mm, magazine-fed, gas-operated, air-cooled, shoulder-fired rifle. On page 3-2, see figure 3-1 for the M16A4 and figure 3-2 for the M4 carbine.

The Service rifle fires in either semiautomatic (i.e., single-shot) mode or in a three-round burst using a selector lever. The M16A4 rifle has a maximum effective range of 550 meters for individual or point targets and the M4 carbine has a maximum effective range of 500 meters. The upper receiver contains the—

- Ejection port.
- Ejection port cover.
- Housing for gas key, bolt-carrier assembly, and bolt assembly.
- Mounting surface for the carrying handle assembly.

The carbine barrel assembly—

- Is air-cooled.
- Contains compensator and front sight assembly.
- Holds the rail system and the sling swivel.

The bore and chamber are chrome-plated to reduce wear and fouling. The rail system protects the gas tube. An aluminum receiver helps reduce the overall weight of the Service rifle. The trigger guard is equipped with a spring-loaded retaining pin that, when depressed, allows the trigger guard to be rotated out of the way for access to the trigger while wearing cold weather gloves. An ejection port cover prevents dirt and sand from getting into the Service rifle through the ejection port. The ejection port cover should be closed when the Service rifle is not being fired. It is automatically opened by the action of the bolt carrier. The muzzle compensator serves as a flash suppressor and assists in reducing muzzle climb.

Figures 3-3, on page 3-2, and 3-4 on page 3-3 present rifle components for the M16A4 rifle and M4 carbine, respectively.

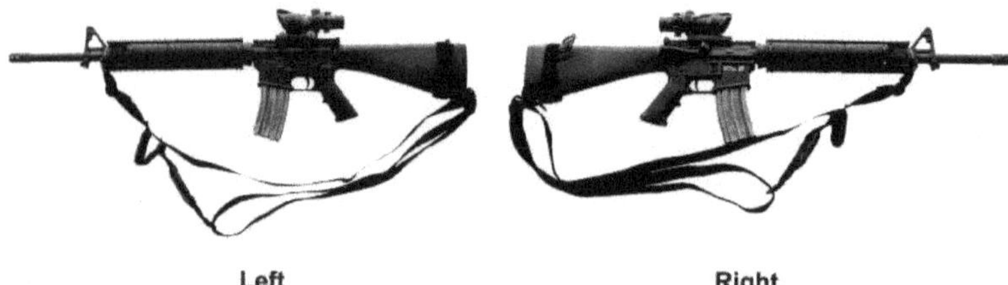

Figure 3-1. M16A4 Rifle.

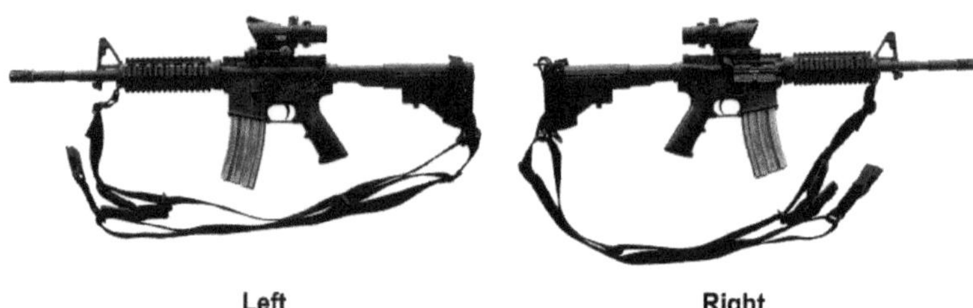

Figure 3-2. M4 Carbine.

Figure 3-3. M16A4 Rifle.

Figure 3-4. M4 Carbine.

The following nomenclature represents the components of the Service rifle:

- *Bolt-carrier assembly.* The bolt-carrier assembly provides stripping, chambering, locking, firing, extraction, and ejection of cartridges using the buffer spring and projectile-propelling gases for power.
- *Charging handle assembly.* The charging handle assembly provides initial charging of the weapon. The handle latch locks the charging handle assembly in the forward position during sustained fire to prevent injury to the operator.
- *Lower receiver and buttstock assembly.* This assembly provides firing control for the Service rifle. The M16A4 also provides storage for basic cleaning materials.
- *Magazine.* The magazine holds cartridges ready for feeding, provides a guide for positioning cartridges for stripping, and provides quick reload capabilities for sustained firing.
- *Sling.* The sling provides the means for carrying the weapon and stability of hold while firing.
- *Upper receiver and barrel assembly.* This assembly provides support for the bolt-carrier assembly. The upper receiver contains the rail system that can accommodate a detachable carrying handle and various other accessories. The barrel chambers the cartridge for firing and directs the projectile.

Operational Controls

The following nomenclature applies to the operational controls of the Service rifle.

Selector Lever

The selector lever has three settings—**SAFE**, **SEMI**, and **BURST**. The firing situation (see fig. 3-5) determines the setting. Modification Instruction 05538/10012A-OR-1A authorizes an ambidextrous selector lever that can be added to the weapon for left-handed shooters only.

Safe. The selector lever in the **SAFE** position prevents the Service rifle from firing.

Semi. The selector lever in the **SEMI** position allows one shot to be fired with each pull of the trigger.

Burst. The selector lever in the **BURST** position allows the Service rifle to continue its cycle of operation until interrupted by the burst cam. With each pull of the trigger, the burst cam limits the maximum number of rounds that can be fired to three. The burst cam is not self-indexing. If burst is selected, the burst cam does not automatically reset to the first shot position of the three-round burst. One, two, or three shots can be fired on the first pull of the trigger. Each subsequent pull of the trigger results in a complete three-round burst unless the trigger is released before the cycle is complete. If the trigger is released during the burst and the three-round cycle is interrupted, the next pull of the trigger fires the rounds remaining in the interrupted three-round cycle.

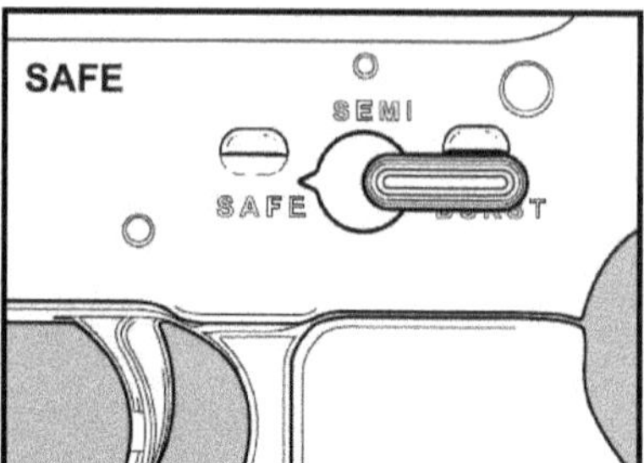

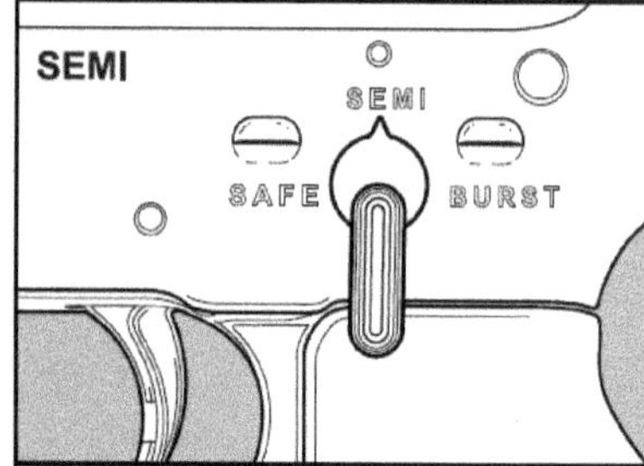

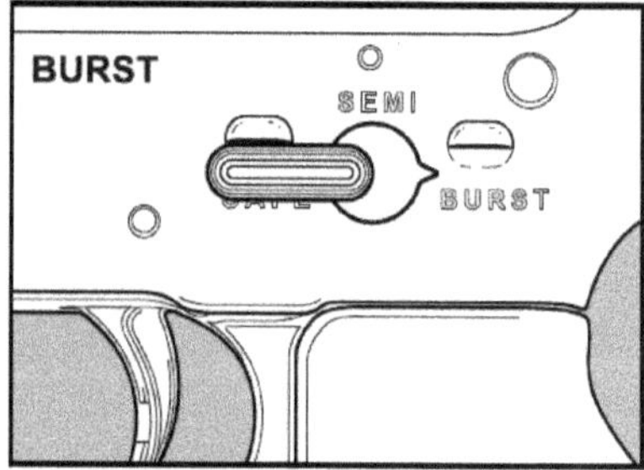

Figure 3-5. Selector Positions: Safe, Semi, and Burst.

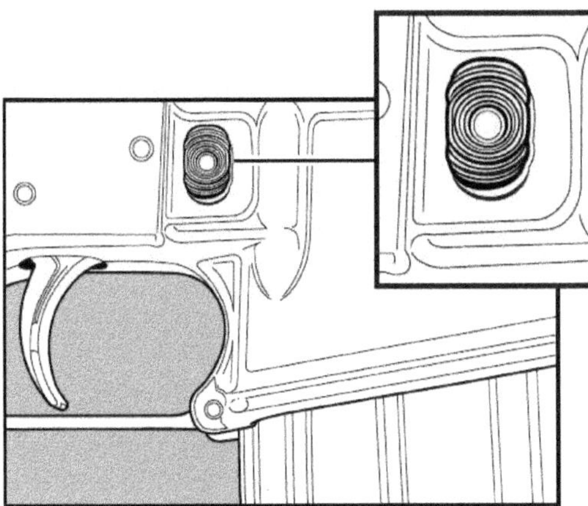

Figure 3-6. Magazine Release Button.

Magazine Release Button

The magazine release button releases the magazine from the magazine well (see fig. 3-6). Modification Instruction 05538/10012A-OR-1A authorizes an ambidextrous magazine release button that can be added to the weapon for left-handed shooters only.

Charging Handle

When the charging handle is pulled to the rear, the bolt unlocks from the barrel extension locking lugs, and the bolt-carrier group moves to the rear of the receiver (see fig. 3-7). Modification Instruction 1005-OR/1 authorizes the optional replacement of the charging handle latch on the Service rifle. The modification consists of an extended charging handle that permits easier manipulation, and a gas deflector, which prevents the gas from blowing into the aiming eye.

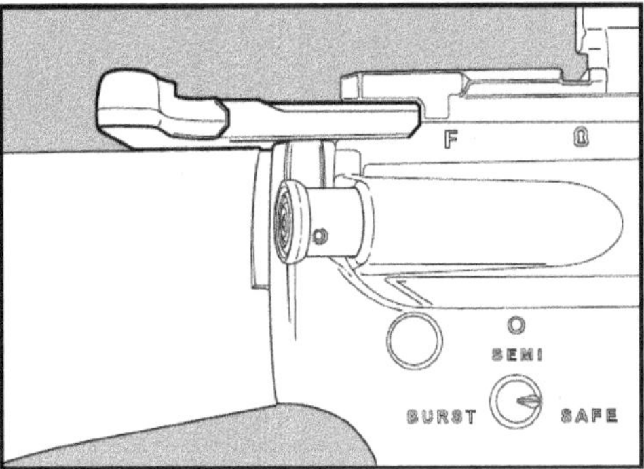

Figure 3-7. Charging Handle.

Bolt Catch

If the charging handle is pulled to the rear when the lower portion of the bolt catch (see fig. 3-8) is depressed, the bolt-carrier group will lock to the rear. When the bolt-carrier group is locked to the rear and the upper portion of the bolt catch is depressed, the bolt-carrier group will slide forward, driven by the buffer assembly and action spring, into the firing position.

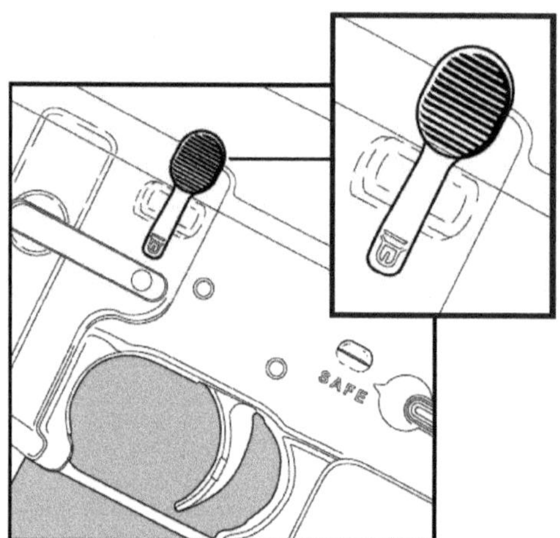

Figure 3-8. Bolt Catch.

Accessories

Rifle Combat Optic AN/PVQ-31A and B

The following applies to the RCO AN/PVQ-31A and B.

Nomenclature and Design

The AN/PVQ-31, known as the RCO, is a compact, dual-source, illuminated telescopic sight. It is a 4 X 32-mm optic that does not require battery power.

The RCO is calibrated to accommodate bullet drop, thereby eliminating the need for mechanical elevation adjustments once a zero has been established. The bullet drop compensator of the AN/PVQ-31A is matched to the trajectory of the M855 5.56-mm round from a 20-inch barrel and supports the M16A4 rifle.

The AN/PVQ-31B is designed for the M4 carbine. Rifle combat optics cannot be switched between different weapon systems. The supported weapon system's designation is inscribed on the inside at the bottom of the optic, and can be read when looking through the lens. Figure 3-9 presents the RCO's controls and indicators.

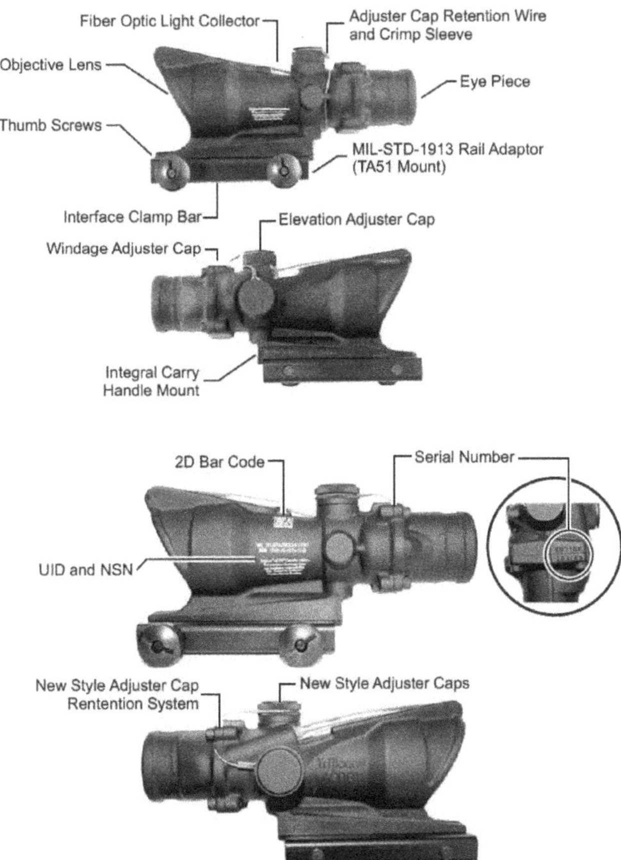

Figure 3-9. Rifle Combat Optic Controls and Indicators.

The reticle pattern consists of a red chevron, horizontal stadia lines, and a left and right horizontal mil scale. The horizontal mil scale is used primarily for communicating target positions and other relationships to team members. The distance from the center post to the first mil bar is 10 mils on each side. The mil scale is graduated in 5-mil increments (see fig. 3-10 on page 3-8).

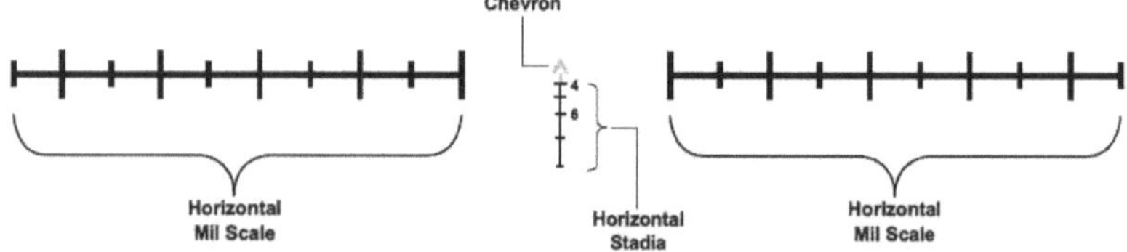

Figure 3-10. Rifle Combat Optic Reticle Pattern.

Methods for Mounting the Rifle Combat Optics

The RCO's TA51 mount is used for mounting the RCO on the M16A4 and M4 rifles (see fig. 3-11) as follows:

- Loosen the two thumb screws, and place the TA51 mount onto the weapon's mounting rail (see fig. 3-12).

> ***Note:*** With the weapon placed into the firing shoulder, have another Marine place the RCO on the mounting rail at a position that supports your eye relief.

Figure 3-11. Rifle Combat Optic Mounted on M16A4.

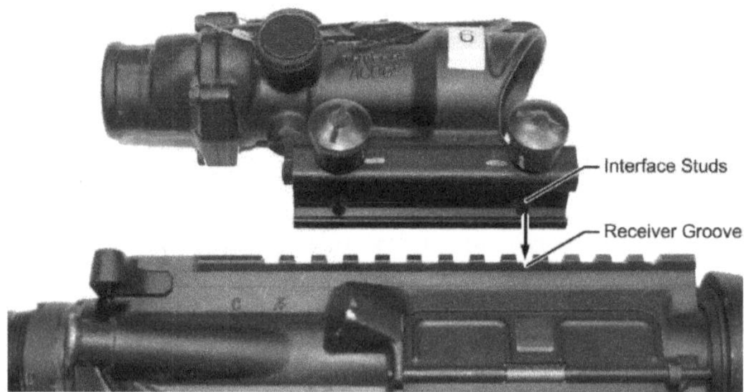

Figure 3-12. Aligning Assembly Mount with Receiver Grooves.

- Push the RCO forward so that the front edge of the recoil stop on the flat top adapter mount is against a recoil groove.
- Tighten the thumb screw on the mount assembly to fix the RCO firmly to the mounting rail.
- Tighten the thumb screws—front and rear at the same time—on the mount assembly to affix the RCO to the rail. Tighten the thumb screws by hand until tight, then tighten a quarter turn more using a tool (e.g., screwdriver).

> *Note:* Based on the operation of the weapon, the TA51 mount must be reversed by an armorer to ensure that the tightening knobs are on the right side of the weapon.

- Check the tightness of the sight and mount during operations and retighten as necessary.

> *Note:* When mounted on the Service rifle, the optic center of the RCO sits approximately 2.75 inches above the centerline of the bore.

Tritium Lamp

The RCO uses a tritium lamp to illuminate the reticle in both low light and complete darkness. The red chevron will also appear illuminated in the daylight. Tritium allows the aiming point to always be illuminated without batteries. It is recommended that the tritium lamp be checked prior to deployment of the optic, every 6 months, or immediately following any incident that might lead to lamp failure (e.g., the dropping of the AN/PVQ-31 onto a hard surface).

> *Note:* To determine that the tritium lamp is functioning in the AN/PVQ-31, enter a dark room and look though the optic. The chevron should be illuminated red. The illumination provided by the tritium lamp is very faint and will be hard to see without eyes that have adapted to the dark. The Marine remains in the dark room for approximately 10 minutes to adapt his eyes to the dark. The reticle is illuminated in low light or complete darkness. If the reticle does not appear to illuminate in the dark, contact the unit maintainer for confirmation and disposal.

MIL-STD-1913 Picatinny Rail System

The rail system (see Military-Standard [MIL-STD]-1913, Dimensioning of Accessory Mounting Rail for Small Arms Weapons) is a bracket used on some firearms to provide a standardized mounting platform for optics and other accessories (e.g., tactical lights, laser sighting modules). The rail system is placed directly on the weapon's receiver in the position normally occupied by the rear sights. Shaped in a cross section roughly like a wide T, optics are mounted on the rail by sliding them on from one end or the other. To provide a stable platform, the rail should not flex as the barrel heats and cools; therefore, most rail systems are cut crosswise, giving them considerable room to expand and contract lengthwise.

The rail can be adapted for use on a multitude of accessories, including optics, flashlights, sights, bipods, M203, and bayonets (with an adaptor). See figure 3-13 on page 3-10.

Figure 3-13. Rail System Accessories.

Vertical Fore Grip (Broomstick)

A fore grip is used to steady firing, counter the effects of recoil, and promote more accurate firing. The fore grip can be incorporated onto some Service rifles as part of the rail integration system (see fig. 3-14).

Figure 3-14. Vertical Fore Grip.

Grip Pod

A grip pod (see fig. 3-15) can be attached to the rail system under the barrel to provide additional stability of hold. The grip pod consists of a vertical grip with extendable, collapsible bipod legs that can be employed in a supported firing position.

Figure 3-15. Grip Pod.

Backup Iron Sight

A backup iron sight (BUIS) may be used instead of the RCO. It is mounted on the rail on the last groove behind the RCO and flips up for use if the RCO becomes disabled. The RCO must be removed to use the BUIS. The sighting system consists of a rear sight aperture, a rear sight elevation drum (range adjuster), and a rear sight windage knob (see fig. 3-16). When mounted on the Service rifle, the rear sight of the BUIS sits approximately 2.75 to 3 inches above the centerline of the bore.

Figure 3-16. Backup Iron Sight.

Rifle Slings

Two rifle slings currently being used by the Marine Corps are the three-point tactical sling and the web sling.

Three-Point Tactical Sling

The application, parts, and nomenclature for the three-point tactical sling are as follows.

Application. Proper application of the three-point tactical sling provides a wide range of mobility when moving. Controlled muscular tension in the firing position is required to provide stability of the sight(s) when employing this sling. The three-point sling is designed mainly for transporting the weapon in a combat application and can be used in the prone, sitting, kneeling, and standing positions with minimal adjustments.

Parts and Nomenclature. The following parts and nomenclature are associated with the three-point tactical sling:

- *Emergency release buckle.* The emergency release buckle allows for quick release of the sling from the body.
- *Flexible swivels with triangular grommet.* There are two front attachments associated with the three-point sling:
  - –The first attachment is designed for use on the Service rifle and is made of Nomex®, which is heat resistant and will prevent melting from the heat of the barrel.
  - –The second attachment is designed for use with various other weapon systems or for use as the rear attachment to the M4 CQBW [close quarter battle weapon]. It is made of nylon and not heat resistant.

Both of the attachments are fastened to the sling by a triangular grommet.

- *1-inch webbing strap with triglide.* The webbing strap allows the sling to be attached to the flexible swivels.
- *Rear stock strap (Cazador).* This strap is used to connect the sling to the buttstock of the weapon.
- *Transition release buckle.* The transition release buckle allows the transition from the strong side to the weak side without removing the sling.

Attachment of the Three-Point Tactical Sling System. To properly attach the three-point tactical sling system, perform the following steps:

- Take the permanently attached front keeper with 1-inch webbing and buckle and feed it through the front, side, or bottom sling mount; then through the buckle.
- Remove the rear stock strap from the sling system and disassemble it by sliding the short end of the rear stock strap off the long end. This should produce an L shape.
- Remove the rear keeper from the triglide. Set the rear keeper with 1-inch webbing aside. It will not be needed.
- Place the weapon with the ejection port cover down and the pistol grip toward you.
- Flip the sling over and lay it flat across the weapon with the buckles facing down. Slide the middle triglide to where it sits on the buttstock, approximately 1/2 inch from the edge of the buttstock. The side of the triglide with one bar should be facing you.
- Place the rear stock strap on the buttstock where the long side is up and the short side is to the right. Feed the short end of the strap through the middle triglide, and pull it to the right until the stitching prevents it from going any further.

> ***Note:*** Left-handed Marines will maintain the same orientation of the rear stock strap. The only difference is that the sling will be pulled tight on the left side of the buttstock where the short strap attaches to the long rear stock strap.

- Flip the sling over so that the quick release buckles are facing up. Wrap the short end of the sling around the back of the buttstock and feed the long end of the stock strap through the slot in the short end.
- Feed the coarse end of the Velcro through the slot closest to the material of the stock strap, then through the other slot on the buckle. Fasten it tightly, ensuring that the triangular grommet is facing down.
- Pull the sling hard to ensure that it is secure.

Three-Point Tactical Sling Positioning/Donning. To achieve proper positioning of the three-point tactical sling, perform the following steps:

- While grasping the pistol grip in your firing hand, place the toe of the Service rifle buttstock in your shoulder.
- While using your support hand, separate the sling with your thumb to create a triangle (see fig. 3-17 on page 3-14).
- While maintaining control of the weapon, insert your head, support hand, and arm into the triangle
- While keeping the sling tight, adjust the triglide so that you can easily bring the weapon into action.

Web Sling

The web sling can be adjusted into either a loop sling or a parade sling. Loop and parade slings are used during training to perfect the application of the fundamentals of marksmanship.

Figure 3-17. Forming a Triangle.

Loop Sling. The loop sling provides the greatest amount of stability during firing and allows the Marine to perfect marksmanship fundamentals. A loop sling takes more time to don; therefore, has limited combat application and is best employed where stability of hold is needed for a precision or long-range shot. The loop sling can be used in the prone, sitting, and kneeling positions.

To achieve proper positioning of the loop sling, perform the following steps:

- Place the Service rifle butt on your firing hip and cradle the Service rifle in your firing arm.
- Disconnect the J-hook from the lower sling swivel.
- With the M-buckle near the hook, feed the sling through the top of the M-buckle to form a loop large enough to slip over your arm (see fig. 3-18).
- Give the loop a one-half turn outboard and insert your support arm through the loop, positioning the loop above your biceps. The loop should be high on your support arm, above your bicep muscle, in such a position that it does not transmit a pulse beat to the Service rifle.
- Position the M-buckle on the outside of your support arm (see fig. 3-19).

Figure 3-18. Forming a Loop.

Figure 3-19. Position of the M-Buckle.

- Tighten the loop on your support arm, ensuring that the M-buckle moves toward the center of your arm as the loop tightens. The sling must pull from the center of your arm to be positioned properly, because as tension is applied to the sling in the firing position, the loop will tighten.
- Loosen the sling keeper and pull up or down (toward or away) from the loop to adjust the sling for the proper length. This adjustment varies with each individual and every firing position. The loop should not be tightened excessively on the arm. If blood flow is restricted, an excessive pulse beat is transmitted through a rifle's sling to the Service rifle and can cause a noticeable, rhythmic movement of the Service rifle sights. When this occurs, a stable hold at the desired aiming point is impossible to achieve.

> *Note:* When the tension being applied on the Service rifle sling is correct, it will cause the butt of the Service rifle to be forced rearward into the pocket of your shoulder. This serves to keep the butt plate in your shoulder pocket during recoil. To increase tension on the Service rifle sling, shorten the sling; and to lessen tension, lengthen the sling.

- Move the sling keeper toward your support arm and secure. The sling keeper should be positioned near the feed end of the sling.
- Place your support hand over the sling from the support side and under the Service rifle. The Service rifle rail system should rest in the V formed between the thumb and forefinger and across the palm of the hand.
- Move your support hand as required to achieve the desired sight picture. Adjust the length of the sling for proper sling tension and support.

See figure 3-20.

Figure 3-20. Loop Sling Donned.

Parade Sling. The parade sling is used to emphasize marksmanship fundamentals while firing from the standing position on known distance (KD) courses during entry level training.

To achieve proper positioning of the parade sling (see fig. 3-21), perform the following steps:

- Attach the sling to the Service rifle by placing the feed end of the sling down through the upper sling swivel.
- Place the feed end of the sling through the sling keeper and lock into place.
- Attach the J-hook to the lower sling swivel.
- Pull the feed end of the sling through the sling keeper until the sling is taut.
- Move the sling keeper down near the feed end of the sling.
- Lock the sling keeper into place.

Figure 3-21. Parade Sling Donned.

Ammunition

The types of ammunition authorized for use with the Service rifle are the ball (M855); tracer (M856); dummy (M199); blank (M200); MK-262 (Mods 0 and 1); and MK-318 Mod 0 (see fig. 3-22 on page 3-18).

> ***Note:*** Refer to the yellow pocket guide *Hazard Classification of United States Military Explosives and Munitions* to identify ammunition by its Department of Defense identification code and other ammunition issued and available in the joint North Atlantic Treaty Organization arena.

Ball (M855)

The ball (M855) is the primary ammunition for the Service rifle. It is identified by a green tip. It has a 62-grain, gilded metal-jacket bullet. The rear two-thirds of the core of the projectile is lead alloy and the front one-third is a solid steel penetrator. The primer and case are waterproofed.

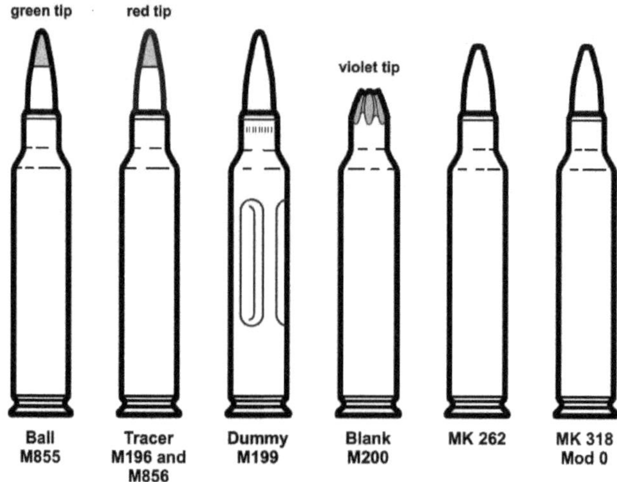

Figure 3-22. Authorized Ammunition.

Tracer (M856)

The tracer (M856) has the same basic characteristics as ball ammunition. It is identified by a bright red tip. Its primary uses include observation firing, incendiary effect, and signaling. Tracer ammunition should be intermixed with ball ammunition in a ratio no greater than 1:1. The preferred ratio is one tracer round to four rounds (1:4).

Dummy (M199)

The dummy (M199) has six grooves along the side of the case. It contains no propellants or primer. The primer well is open to prevent damage to the firing pin. The dummy cartridge is used during dry fire and other training applications.

Blank (M200)

The blank (M200) has the case mouth closed with a seven-petal rosette crimp. It contains no projectile. Blank ammunition, identified by its violet tip, is used for training purposes.

MK-262

The MK-262 is a 5.56-mm centerfire cartridge. It is a 77-grain open-tipped match bullet. It has no color coding on the projectile.

MK-318 Mod 0

This SOST [Special Operations Science and Technology] 5.56-mm ammunition is a 62-grain bullet. It features an open tip with lead at the front, a thick copper base, and has no color coding on the projectile. The round is designed to be "barrier blind," meaning that the round stays on target better than the existing M855 round after penetrating windshields, car doors, and other objects.

Internal Ballistics

Ballistics is the science of projectiles and their effects. Internal ballistics is what a projectile is doing while it is inside a weapon. As the powder burns, chamber pressure builds, and the projectile is pushed forward through the throat, engaging the rifling (see fig. 3-23).

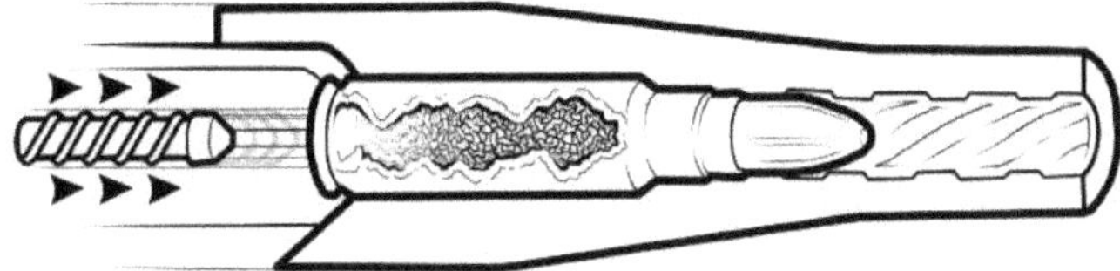

Figure 3-23. Chamber Pressure.

The powder in the cartridge continues to burn, adding pressure and increasing the velocity of the projectile (see fig. 3-24).

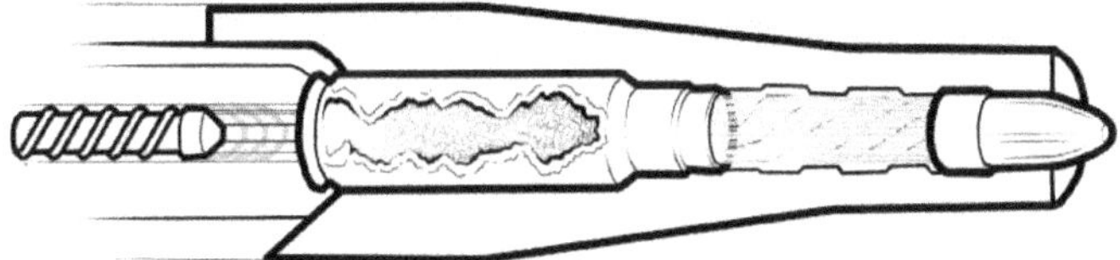

Figure 3-24. Velocity of Projectile.

As the round travels down the barrel, it is being rotated by the rifling cut into the barrel. This rotation stabilizes the round once it exits the barrel. The stability of the round is determined by the twist rate and is designed to give the optimal amount of stability to a given projectile (see fig. 3-25). For optimal performance, the longer, heavier rounds require more spin than the lighter, shorter rounds. If the twist rate is incorrect, it will adversely affect the round. If the round is over-stabilized, the projectile will perform poorly when it impacts body tissue, punching straight through with minimal disturbance of the tissue. If the projectile is not stabilized enough it will not travel accurately and will be heavily affected by atmospheric conditions and drag.

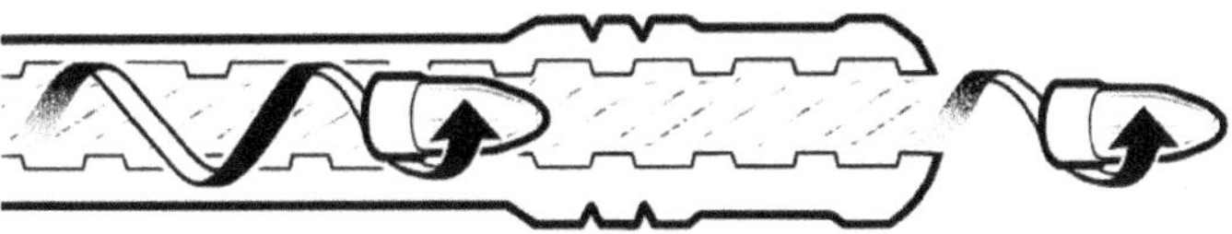

Figure 3-25. Round Exiting Barrel.

Cycle of Operation

The cycle of operation includes firing, unlocking, extracting, ejecting, cocking, feeding, chambering, and locking.

Firing

The hammer releases and strikes the head of the firing pin, driving the firing pin into the round's primer. The primer ignites the powder in the cartridge. Gas generated by the rapid burning of powder propels the projectile through the barrel. After the projectile passes the gas port, a portion of the expanding gas enters the gas port and gas tube. The gas tube directs the gas rearward into the bolt-carrier key and causes the bolt carrier to move rearward. See figure 3-26.

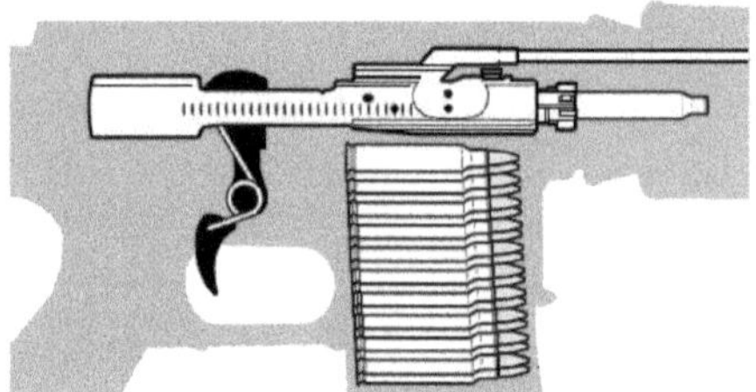

Figure 3-26. Firing.

Unlocking

Figure 3-27 illustrates unlocking of the bolt. As the bolt carrier moves to the rear, the bolt cam pin follows the path of the cam track located in the bolt carrier. This causes the bolt assembly to rotate until the bolt-locking lugs are no longer aligned behind the barrel extension locking lugs.

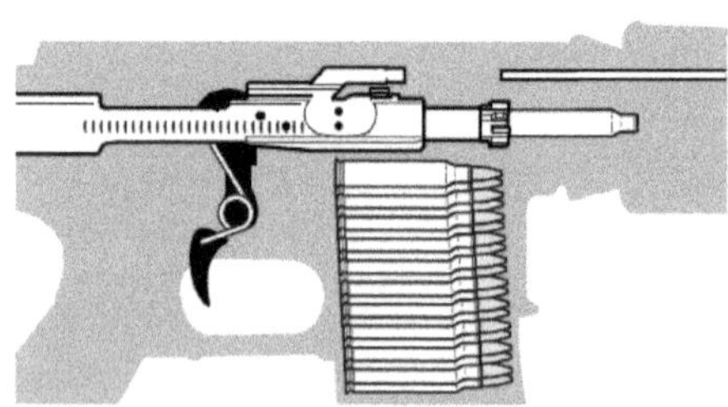

Figure 3-27. Unlocking.

Extracting

As the bolt-carrier group continues to move to the rear, the extractor claw withdraws the cartridge case from the chamber. See figure 3-28.

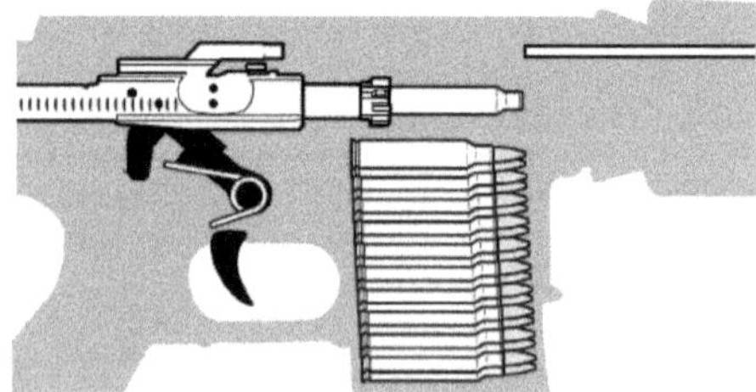

Figure 3-28. Extracting.

Figure 3-29. Ejecting.

Ejecting

The ejector, located in the bolt face, is compressed into the bolt body by the base of the cartridge case. The rearward movement of the bolt-carrier group allows the nose of the cartridge case to clear the front of the ejection port. The cartridge case is thrown out by the action of the ejector and spring. See figure 3-29.

Cocking

Continuing its rearward travel, the bolt carrier overrides the hammer, forces it down into the receiver, compresses the hammer spring, and causes the disconnector to engage the lower hammer hook. See figure 3-30.

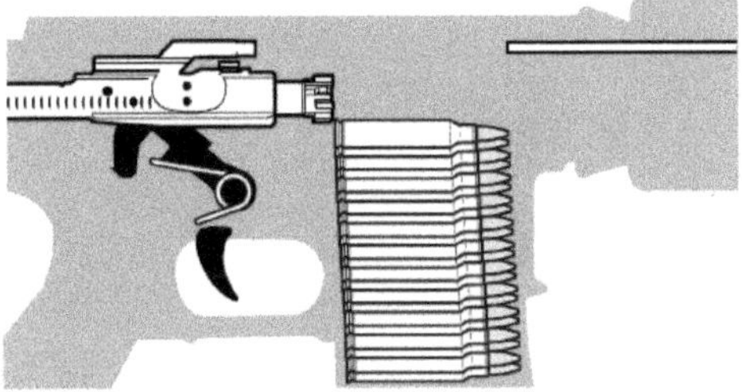

Figure 3-30. Cocking.

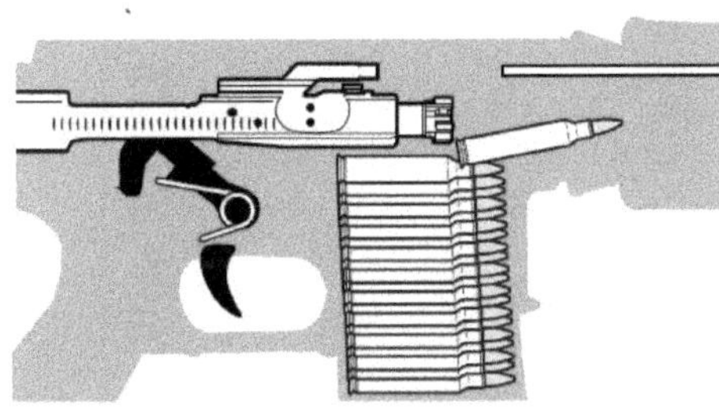

Figure 3-31. Feeding.

Feeding

Once the rearward motion causes the bolt-carrier group to clear the top of the magazine, the expansion of the magazine spring forces a round into the path of the bolt. After the action spring overcomes and absorbs the rearward motion of the bolt-carrier group, it expands and sends the buffer assembly and bolt-carrier group forward with enough force to strip a round from the magazine. See figure 3-31.

Chambering

As the bolt-carrier group continues to move forward, pushing a fresh round in front of it, the face of the bolt thrusts the new round into the chamber. The extractor claw grips the rim of the cartridge case. The ejector is forced into its hole, compressing the ejector spring. See figure 3-32.

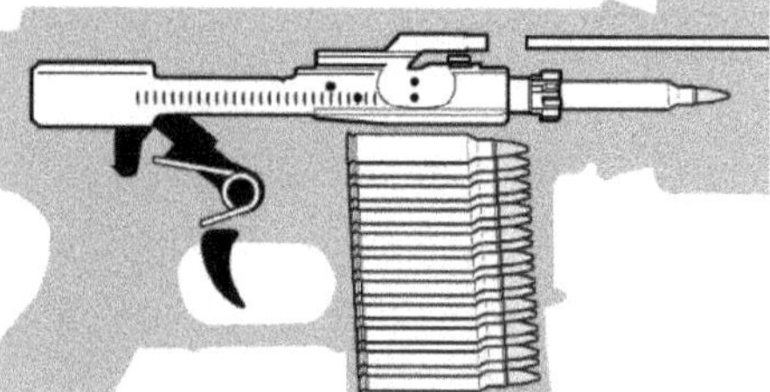

Figure 3-32. Chambering.

Locking

As the bolt-carrier group continues to move forward, the bolt-locking lugs are forced against the barrel extension, and the bolt cam pin is forced along the cam track. The bolt rotates and aligns the bolt-locking lugs behind the barrel extension locking lugs. The weapon is ready to fire. See figure 3-33.

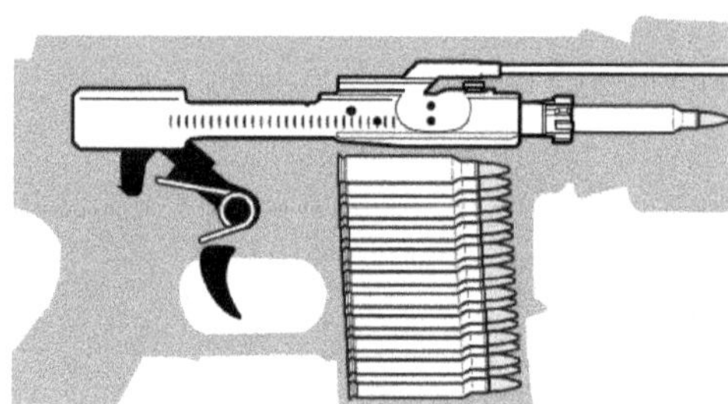

Figure 3-33 Locking.

Preventive Maintenance

Normal care and cleaning of the Service rifle will result in the proper functioning of all parts. Service rifle care and cleaning involves proper disassembly, cleaning, and reassembly. Only the proper issue-type of cleaning materials can be used. Improper maintenance can cause stoppages, reducing combat readiness and effectiveness.

Main Group Disassembly

The Service rifle is disassembled into three main groups—upper receiver, bolt carrier, and lower receiver (see fig. 3-34). Before disassembling the Service rifle—

- Ensure that the weapon is in Condition 4 (see pg. 4-2).
- Remove the sling.

Upper Receiver

To disassemble the upper receiver—

- Move the take-down pin from the left to the right as far as it will go to allow the lower receiver to pivot down from the upper receiver.
- Move the receiver pivot pin from the left to the right as far as it will go and separate the upper and lower receivers.

- Pull back the charging handle and bolt carrier approximately 3 inches and remove the bolt-carrier group.
- Remove the charging handle by sliding it back and down, out of the upper receiver.

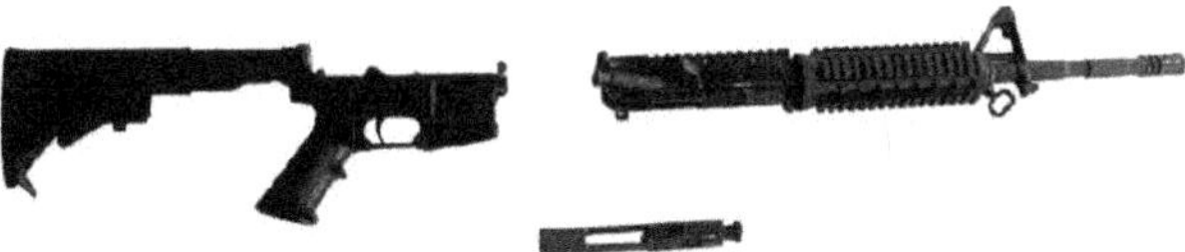

Figure 3-34. Three Main Groups.

Bolt Carrier

To disassemble the bolt carrier—

- Remove the firing pin retaining pin.
- Push the bolt back into the bolt carrier to the locked position.
- Tap the base of the bolt carrier against the palm of your hand so that the firing pin will drop out.
- Rotate the bolt cam pin one-quarter turn, and lift the bolt cam pin out.
- Withdraw the bolt assembly from the carrier.
- Press on the extractor's rear and use the firing pin to push out the extractor-retaining pin. Remove the extractor and spring (the spring is permanently attached to the extractor).

See figure 3-35.

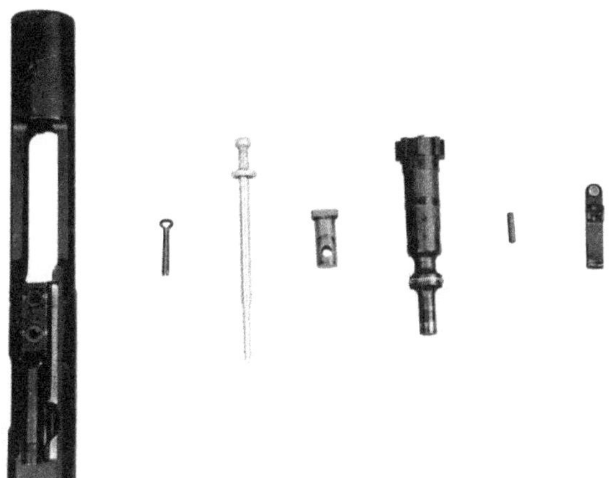

Figure 3-35. Bolt Carrier Disassembled.

CAUTION

Be careful not to damage the tip of the firing pin while pushing out the extractor-retaining pin.

Note: The extractor assembly has a rubber insert within the spring. Do not attempt to remove it. If the spring comes loose, put the large end of the spring in the extractor and seat it. Push in the extractor pin.

Lower Receiver

To disassemble the lower receiver—

- Press in the buffer and depress the buffer retainer.

Note: It may be necessary to use the edge of the charging handle to depress the buffer retainer.

- Press the hammer downward and ease the buffer and action spring forward and out of the receiver. Separate the parts.

See figure 3-36. No further disassembly of the lower receiver is required.

Figure 3-36. Lower Receiver Disasesmbled.

Note: In combat situations, the Service rifle may be partially disassembled in any sequence; however, combat situations are the exception, not the rule. Under normal operating circumstances, disassemble the Service rifle in the sequence described beginning on page 3-22. Any further disassembly of the Service rifle should be performed by a qualified armorer.

Magazine Disassembly

The magazine should be disassembled regularly for cleaning to avoid the possibility of malfunction or stoppage of the Service rifle caused by dirty or damaged magazines. To disassemble the magazine—

- Pry up and push the base plate out from the magazine.
- Jiggle the spring and follower to remove. Do not remove the follower from the spring.

See figure 3-37.

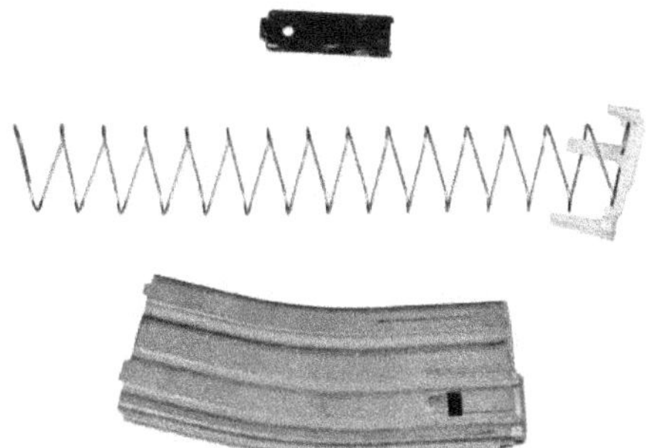

Figure 3-37. Magazine Disassembled.

Proper Cleaning

Cleaning Materials

In accordance with Marine Corps stock list, SL-3-11290_, the following cleaning materials are used for preventive maintenance (see fig. 3-38 on page 3-26).

- *Cleaner, lubricant, and preservative.* When using cleaner, lubricant, and preservative (CLP), always shake the bottle well before each use. In all but the coldest arctic conditions, CLP is the lubricant that should be used for the Service rifle.
- *Rod.* The rod is in three sections, with a handle assembly.
- *Brushes.* The brushes are bore, chamber, and general purpose.
- *Patch holder.* The patch holder includes swabs, patches, pipe cleaners, and clean rags.

Cleaning the Upper Receiver

Basic cleaning of the upper-receiver group should include the following:

- Attach the three rod sections together, but leave each one approximately two turns short of being tight.
- Attach the patch holder onto the rod.

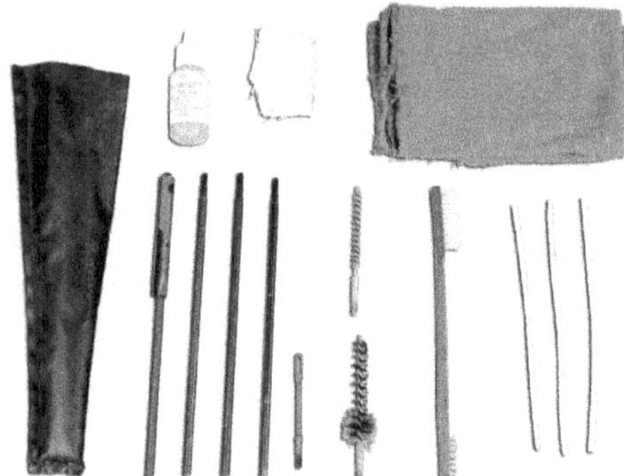

Figure 3-38. Cleaning Materials.

- Point the muzzle down and insert the nonpatch end of the rod into the chamber. Attach the handle to the cleaning rod section and pull a CLP-moistened, 5.56-mm patch through the bore.
- Attach the bore brush to the rod but leave it two turns short of being tight, then—
  –Put a few drops of CLP on the bore brush.
  –Insert the rod into the barrel from the chamber end, attach the handle, and pull the brush through the bore. Repeat three times.
  –Remove the bore brush and attach the patch holder to the rod with a CLP-moistened patch, insert the rod into the barrel from the chamber end, attach the handle, and pull the patch through the bore.
- Inspect the bore for cleanliness by holding the muzzle to your eye and looking into the bore.
- Repeat the steps for cleaning the upper receiver until the patches come out of the bore clean.
- Attach the chamber brush and one section of the cleaning rod to the handle. Moisten it well with CLP and insert it into the chamber.
- Scrub the chamber and bolt lugs using a combination of a plunging and clockwise rotating action.

Note: Do not reverse direction of the brush while it is in the chamber.

- Clean the interior portion of the upper receiver with the general-purpose brush and CLP.
- Dry the bore, chamber, and the interior of the receiver with rifle patches, swabs, and clean rags until they come out clean. Then moisten all interior surfaces with CLP.
- Wipe the barrel, gas tube, and rail system clean with a rag.

Cleaning the Bolt-Carrier Group

When cleaning the bolt-carrier group—

- Clean the outer and inner surfaces of the bolt carrier with a general-purpose brush.
- Clean the bolt-carrier key with a pipe cleaner.

- Clean the locking lugs, gas rings, and exterior of the bolt with the general-purpose brush.
- Insert a swab into the rear of the bolt and swab out the firing pin recess and gas ports.
- Clean the extractor groove with the general-purpose brush, ensuring all the carbon is removed from underneath the extractor lip.
- Clean extractor pin, firing pin, and firing pin retaining pin, using the general-purpose brush and CLP.
- Clean the charging handle assembly with the general-purpose brush and patches.

Cleaning the Lower Receiver

When cleaning the lower receiver—

- Wipe dirt from the firing mechanism using a general-purpose brush, clean patch, pipe cleaners, and swabs.
- Clean the outside of the receiver with the general-purpose brush and CLP. Clean the buttplate and rear sling swivel, ensuring that the drain hole is clear of dirt.
- Wipe the inside of the buffer tube, buffer, and action spring.
- Wipe the inside of the magazine well with a rag.
- Wipe out the inside of the pistol grip, and ensure that it is clean.

Cleaning the Magazine

When cleaning the magazine—

- Clean the inside of the magazine with the general-purpose brush and CLP.
- Wipe the magazine dry.
- Keep the spring lightly oiled.

Inspection

While cleaning the Service rifle, and during each succeeding step in the preventive maintenance process, inspect each part for cracks and chips and ensure that parts are not bent or badly worn. Report any damaged part to the armorer.

Inspect the carrying handle assembly and mounting surface of the upper receiver for damage. If the carrying handle is missing or cannot be correctly mounted, repair as authorized, or evacuate to support maintenance.

Inspect the carrying handle assembly to ensure that the unit-applied identification code matches the unit-applied identification code on the weapon. If it does not match, locate the correct carrying handle assembly, and match it to the correct Service rifle. If a match cannot be found, the weapon should be rezeroed by the operator.

Inspection is a critical step to ensure the combat readiness of the Service rifle. Normally, it is performed during rifle cleaning (prior to lubrication); however, it can be performed throughout the preventive maintenance process.

Lubrication

Lubrication is performed as part of the detailed procedure for preventive maintenance and in preparation for firing.

> *Note:* Lightly lubricate means that a film of CLP barely visible to the eye should be applied. Generously lubricate means that the CLP should be applied heavily enough that it can be spread with a finger.

Upper Receiver

To lubricate the upper receiver—

- Lightly lubricate the inside of the upper receiver, bore, chamber, outer surfaces of the barrel.
- Depress the front sight detent, apply two or three drops of CLP, and depress several times to work the lubrication into the spring.
- Lubricate the moving parts and elevation screw shaft of the rear sight.
- Remove excess CLP from the bore and chamber before firing.

Bolt-Carrier Group

To lubricate the bolt-carrier group—

- Generously lubricate the outside of the cam pin area, the bolt rings, and outside of the bolt body.
- Lightly lubricate the charging handle and the inner and outer surfaces of the bolt carrier.

Lower Receiver

To lubricate the lower receiver—

- Lightly lubricate the inside of the lower-receiver extension.
- Generously lubricate the moving parts, including any pins in those parts, located inside the lower receiver.

Reassembly

Reassembling the Service Rifle

When reassembling the Service rifle—

- Return all cleaning gear.
- Connect the buffer and action spring and insert them into the buffer tube/stock.
- Place the extractor and spring back on the bolt. Depress the extractor to align the holes and reinsert the extractor pin.
- Insert the bolt into the carrier.

Warning: Do not switch bolts between Service rifles because a catastrophic failure could occur.

- Hold the bolt carrier with the bolt-carrier key at 12 o'clock. Insert the bolt into the bolt carrier with the extractor at 12 o'clock.
- Rotate the bolt counterclockwise until the cam pinhole aligns to the cam pin slot in the bolt carrier.

> **WARNING**
>
> Ensure that the cam pin is installed in the bolt group or the Service rifle may explode while firing.

- Insert the bolt cam pin through the bolt carrier and into the bolt. Rotate the cam pin one-quarter turn to either the right or left. Pull the bolt forward until it stops.
- Drop in the firing pin from the rear of the bolt carrier and seat it.
- Replace the firing pin retaining pin. Ensure that the head of the firing pin retaining pin is recessed inside the bolt carrier. The firing pin should not fall out when the bolt-carrier group is turned upside down.
- Place the charging handle in the upper receiver by lining it up with the grooves in the receiver. Push the charging handle in partially.
- With the bolt in the unlocked position, place the bolt-carrier key into the groove of the charging handle. Push the charging handle and bolt-carrier group into the upper receiver until the charging handle locks. Join the upper and lower receivers and engage the receiver pivot pin.
- Ensure that the selector lever is on **SAFE** before closing the upper receiver. Close the upper and lower receiver groups. Push in the take-down pin.
- Install the rail covers.
- Attach the sling.
- Place the weapon in Condition 4 (see pg. 4-2). Pull the charging handle to the rear and release. Ensure that the selector lever is on **SAFE**, and pull the trigger. The hammer should not fall.
- Place the selector lever on **SEMI**. Pull the trigger and hold it to the rear. The hammer should fall. Pull the charging handle to the rear and release. Release the trigger and pull again. The hammer should fall.
- Pull the charging handle to the rear and release. Place the selector lever on **BURST**. Pull the trigger, and hold it to the rear. The hammer should fall. Pull the charging handle to the rear three times and release. Release the trigger and pull again. The hammer should fall.
- Pull the charging handle to the rear and release. Place the selector lever on **SAFE**.

Reassembling the Magazine

To reassemble the magazine—

- Insert the follower and jiggle the spring to install.
- Slide the base under all four tabs until the base catches.
- Ensure that the printing is on the outside.

Field Maintenance

Preventive maintenance in the field is performed when detailed disassembly and cleaning is not practical due to operational tempo or the level of threat. To perform limited field preventive maintenance—

- Place the Service rifle in Condition 4 (see pg. 4-2).
- Break the Service rifle down by removing the rear take-down pin and rotating the upper receiver and barrel forward.
- Remove the bolt-carrier group.

Note: Do not disassemble the bolt-carrier group further.

- Clean the bolt-carrier group as follows:
  –Clean the upper and lower-receiver groups without further disassembly.
  –Clean the bore and chamber. Lubricate the Service rifle.
  –Reassemble the Service rifle and perform a user-serviceability inspection (see pgs. 3-29 and 3-32, respectively).

Service Rifle Cleaning in Various Climatic Conditions

The climatic conditions in various locations require special knowledge about cleaning and maintaining the Service rifle. The conditions that will affect the Service rifle include hot, wet, tropical; hot, dry desert; arctic or low temperature; and heavy rain and fording.

Hot, Wet, Tropical

During hot, wet, tropical conditions—

- Perform normal maintenance.
- Clean and lubricate the Service rifle more often. Inspect hidden surfaces for corrosion. Pay particular attention to spring-loaded detents.
- Use lubricant more liberally.
- Empty and check the inside of the magazine more frequently. Wipe dry and check for corrosion.
- Keep the Service rifle covered, when practical.

Hot, Dry Desert

Hot dry climates are usually areas that contain blowing sand and fine dust. Dust and sand will get into the Service rifle and magazines, causing stoppages. It is imperative to pay particular attention to the cleaning and lubrication of the Service rifle in this type of climate. Dry, dusty climates require more frequent maintenance and the Service rifle must be lubricated frequently.

Arctic or Low Temperature

During arctic or low temperature conditions—

- Clean and lubricate the Service rifle in a warm room, with the Service rifle at room temperature, if possible. Lubricating oil, arctic weapons can be used below a temperature of 0 degrees Fahrenheit and must be used below -35 degrees Fahrenheit.
- Keep the Service rifle covered when moving from a warm to a cold environment, allowing gradual cooling of the Service rifle. This prevents condensation of moisture and subsequent freezing.
- Leave the Service rifle in a protected, but cold area outdoors, if possible. When bringing the Service rifle inside to a warm place, it should be disassembled and wiped down several times as it warms, because condensation will form on the Service rifle when it is moved from outdoors to indoors.
- Keep the Service rifle dry when possible.
- Unload and perform a function check every 30 minutes, if possible, to help prevent freezing of functional parts.
- Do not lay a warm Service rifle in snow or ice.
- Keep the inside of the magazine and ammunition wiped dry. Moisture will freeze and cause stoppages.

Heavy Rain and Fording

During heavy rains and fording conditions—

- Keep the Service rifle dry and covered when practical.
- Keep water out of the barrel if possible. If water does get into the Service rifle, drain and dry with a patch, if available. If water is in the barrel, point the muzzle down and break the seal by doing a chamber check so that the water will drain. If water is in the stock of the weapon, ensure that the drain hole in the stock is clear so the water can run out.
- Perform normal maintenance.

Function Check

A function check is performed to ensure that the Service rifle operates properly after the weapon has been reassembled. To perform a function check—

- Ensure that the—

  –Magazine is removed.

  –Chamber is empty.

  –Bolt is forward.

  –Safety is on.

  –Ejection port cover is closed.

- Pull the charging handle to the rear and release. Ensure that the selector lever is on **SAFE** and pull the trigger. The hammer should not fall.
- Place the selector lever on **SEMI**. Pull the trigger and hold it to the rear. The hammer should fall. Pull the charging handle to the rear and release. Release the trigger and pull again. The hammer should fall.
- Pull the charging handle to the rear and release.
- Place the selector lever on **BURST**. Pull the trigger and hold it to the rear. The hammer should fall. Pull the charging handle to the rear three times and release. Release the trigger and pull again. The hammer should fall.
- Pull the charging handle to the rear and release. Place the selector lever on **SAFE**.

User-Serviceability Inspection

Individual Marines must perform user-serviceability inspections on their weapons before firing them. This inspection ensures—

- That the weapon is in an acceptable operating condition.
- That an inspection is performed prior to any combat operation (e.g., a patrol) being posted.

> *Note:* This inspection does not replace a limited technical inspection or prefire inspection conducted by a qualified armorer.

To perform a user-serviceability inspection, perform the following steps:

- Place the Service rifle in Condition 4 (see pg. 4-2).
- Conduct a function check.
- Check the Service rifle to ensure that the—

  –Compensator is tight and oriented correctly, vents up.

  –Barrel is tight.

  –Barrel is clear of obstructions.

  –Front sight post is straight and adjustable.

  –Rails and rail covers are serviceable. Any uncovered portions of the rail are protected with rail covers. Ensure that the screw on the top aft portion of the rail is present and tight.

  –RCO is securely mounted to the rail system, that there are no cracks or scratches on the optic lens, and the—

  - Fiber optic light collector is serviceable.
  - Protective caps are present and secure.
  - Screws that secure the RCO to the rail are secure.
  - Thumbscrews on the RCO are mounted over the ejection port cover.

  –Gas tube is securely fixed in the front sight assembly.

  –Gas tube, from the chamber end, is not bent or damaged and is in uniform shape.

  –Bolt-carrier group is properly assembled, rotates freely, and that the gas rings are staggered evenly around the bolt.

–Magazine releases.

–Pistol grip is tight.

–Trigger guard release pin is locked into place.

–Ejection port operates correctly.

–Rear sight elevation and windage knobs are adjustable and have distinct clicks.

–Stock is tight on the lower receiver.

–Buffer tube is straight and not cracked.

–Weapon is properly lubricated for operational conditions.

–Sling keeper can be adjusted and secured.

- Ensure that the magazines are serviceable and clean by performing the following steps:

  –Load the Service rifle with an empty magazine. Ensure that the magazine can be seated.

  –Pull the charging handle to the rear without depressing the bolt catch. Ensure that the bolt locks to the rear.

  –Depress the upper portion of the bolt catch and observe the bolt moving forward on an empty chamber. Ensure the bolt moves completely forward and locks in place.

  –Repeat this procedure with all magazines.

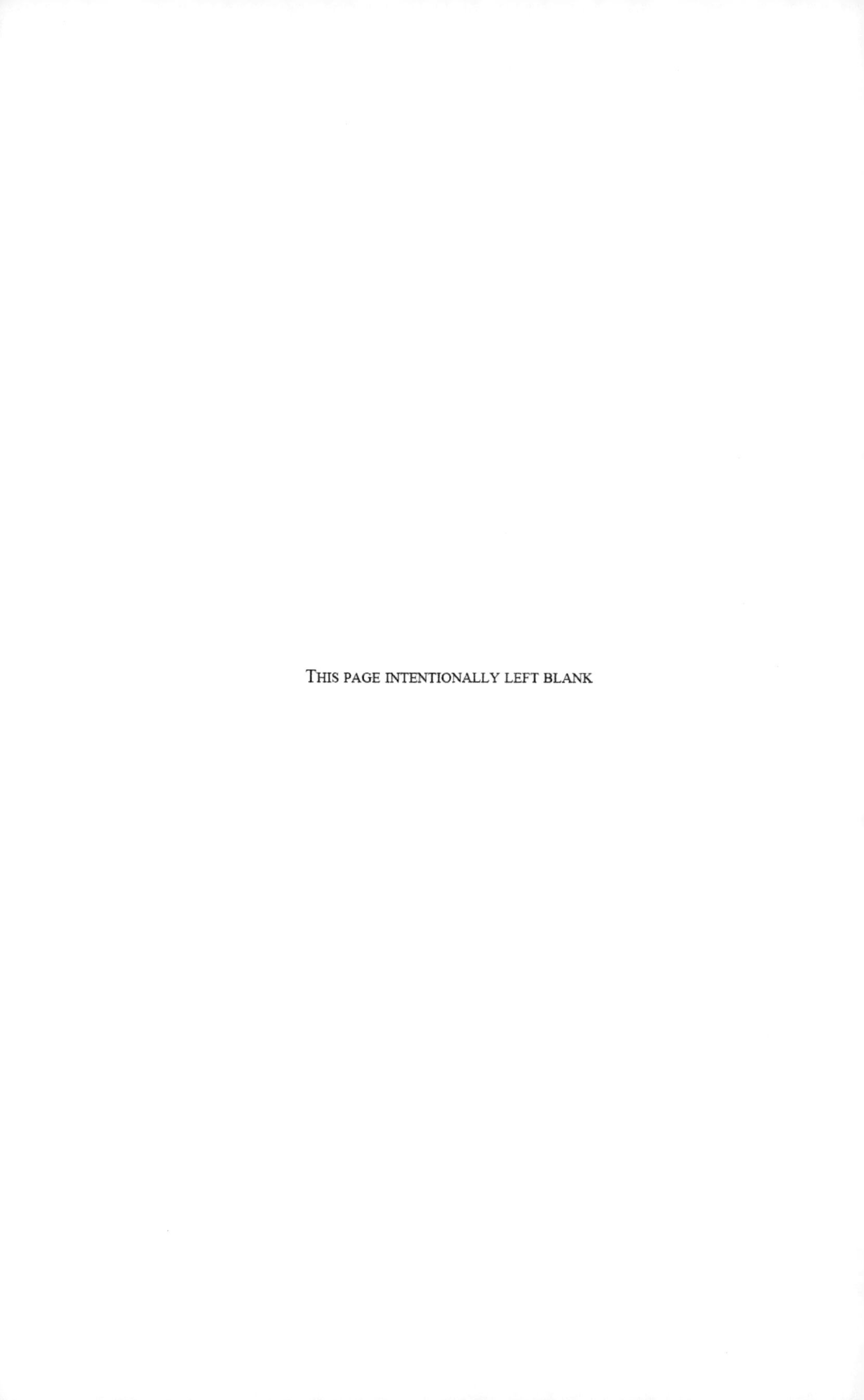

THIS PAGE INTENTIONALLY LEFT BLANK

CHAPTER 4

Weapons Handling

Weapons handling procedures provide a consistent and standardized way for Marines to handle, operate, and employ the Service rifle safely and effectively. Proper weapons handling procedures—

- Ensure the safety of Marines by eliminating negligent discharges and reinforcing positive identification of targets before engagement.
- Apply at all levels of training and during combat operations.

Safety Rules

Safe Service rifle handling is critical. If proper weapons handling procedures are not observed the Marine risks both his safety and the safety of his fellow Marines. During combat, the Marine must react quickly, safely, and be mentally prepared to engage targets. To ensure that only the intended target is engaged, the Marine must apply the following safety rules at all times:

Rule 1: Treat every weapon as if it were loaded.
When the Marine takes charge of a Service rifle in any situation, he must treat the weapon as if it were loaded, determine its condition, and continue applying the other safety rules.

Rule 2: Never point a weapon at anything you do not intend to shoot.
The Marine must maintain muzzle awareness at all times.

Rule 3: Keep your finger straight and off the trigger until you are ready to fire.
A target must be identified before moving your finger to the trigger.

Rule 4: Keep the weapon on SAFE until you intend to fire.
A target must be identified before taking the weapon off SAFE.

Weapons Conditions

A weapon's readiness is described by conditions. The steps involved in the loading and unloading process take the Service rifle through the specific conditions of readiness for live fire.

Condition 1. Safety on, magazine inserted, round in chamber, bolt forward, and ejection port cover closed.

Condition 2. Not applicable to the Service rifle.

Condition 3. Safety on, magazine inserted, chamber empty, bolt forward, and ejection port cover closed.

Condition 4. Safety on, magazine removed, chamber empty, bolt forward, and ejection port cover closed.

Weapons Condition Determination

When the Marine takes charge of a weapon in any situation, its condition must be determined and known at all times. Situations may include the following:

- Discovering an unmanned Service rifle in combat.
- Taking charge of any weapon after it has been unmanned (e.g., out of a rifle rack, stored in a vehicle).
- Taking charge of another Marine's weapon.

To determine the condition of the weapon in any of these situations, the Marine must—

- Determine if a magazine is present.
- Ensure that the weapon is on **SAFE**.
- Conduct a chamber check.

To conduct a chamber check—

- Bring your support hand back against the magazine well so that the slip ring rests in the V of the hand.
- Extend the fingers of your support hand and cover the ejection port (right-handed Marine) or extend your thumb over the ejection port (left-handed Marine).
- Grasp the charging handle with your index and middle fingers of your firing hand. Control the weapon by pointing the muzzle to the deck and applying tension against the stock with the heel of your hand.
- Pull the charging handle slightly to the rear and visually and physically inspect the chamber. For right-handed Marines, insert one finger of the support hand into the ejection port and feel whether a round is present (see fig. 4-1). For left-handed Marines, insert the thumb of the right hand into the ejection port and feel whether a round is present (see fig. 4-2).
- Release the charging handle and observe the bolt going forward.

Figure 4-1. Right-Handed Chamber Check.

Figure 4-2. Left-Handed Chamber Check.

CAUTION

Pulling the charging handle too far to the rear while inspecting the chamber can cause double feed or ejection of one round of ammunition.

- Tap the forward assist.
- Close the ejection port cover if time and the situation permits.
- Remove the magazine if one is present and observe if ammunition is present. If time permits, count the rounds. Reinsert the magazine into the magazine well.

Note: The same procedure is used during both daylight and low visibility. A chamber check may be conducted at any time.

Weapons Commands

Weapons commands dictate the specific steps required to load and unload the Service rifle. The following commands are used when handling weapons:

- On the command *make a condition 3 weapon*, load the weapon, taking it from Condition 4 to Condition 3 by inserting a filled magazine.
- On the command *make a condition 1 weapon*, make the weapon ready, taking it from Condition 3 to Condition 1 by chambering a round.
- The command *fire* is used to specify when the Marine may engage targets.
- The command *cease fire* is used to specify when the Marine must stop target engagement.
- On the command *make a condition 4 weapon*, unload the weapon, taking it from any condition to Condition 4 (see pg. 4-2).
- The command *show clear* is used when a secondary observation is required to verify that no ammunition is present before the Service rifle is placed in Condition 4 (see pg. 4-2).

Load

On the command *make a condition 3 weapon*, the Service rifle is loaded as follows:

- Ensure that the Service rifle is on **SAFE**.
- Withdraw the magazine from the magazine pouch that is farthest to reach with your support hand.
- Observe the magazine to ensure that it is filled.
- Insert the magazine fully into the magazine well. The magazine catch will click as it engages and can be felt and/or heard by the Marine. Without releasing the magazine, tug downward on the magazine to ensure that it is seated.
- Close the ejection port cover.
- Fasten the magazine pouch.

Make Ready

On the command *make a condition 1 weapon*, the Service rifle is made ready as follows:

- Pull the charging handle to its rearmost position and release. Grip the pistol grip with your firing hand and pull the charging handle with your support hand (see fig. 4-3).

  Note: When employing a loop sling, grasp the rail system with your support hand and pull the charging handle with your firing hand.

- Conduct a chamber check (see pg. 4-2) to ensure that a round has been chambered.

Figure 4-3. Pulling Charging Handle with Support Hand.

- Close the ejection port cover if time and the situation permit.
- Check the sights to ensure proper battlesight zero (BZO) setting and correct rear sight aperture.

Note: This only applies to iron sights.

Fire

On the command *fire*—

- Aim the Service rifle.
- Take the Service rifle off **SAFE**.
- Press the trigger to the rear.

Cease Fire

On the command *cease fire*—

- Place your trigger finger straight along the receiver.
- Place the weapon on **SAFE**.

Unload

On the command *make a condition 4 weapon*, the Service rifle is unloaded as follows:

- Ensure that the weapon is on **SAFE**.
- Remove the magazine from the Service rifle and retain it on your person.
- Bring your support hand back against the magazine well so that the slip ring rests in the V of the hand. While cupping your support hand under the ejection port, rotate the Service rifle so that the ejection port is facing the deck.
- Pull the charging handle to the rear and catch the round in your support hand (see fig. 4-4).

Figure 4-4. Catching the Round.

Note: If there is a slight indentation on the primer, it does not indicate grade 3 ammunition. This normally occurs because of the inertia of the bolt chambering the round.

- Lock the bolt to the rear.
- Put the weapon on **SAFE** if the selector lever would not move to **SAFE** earlier.
- Ensure that the chamber is empty and no ammunition is present.
- Depress until the bolt catches, then observe that the bolt is moving forward on an empty chamber (see fig. 4-5).

Figure 4-5. Observing the Chamber.

- Close the ejection port cover.
- Check the sights. For proper BZO setting, correct rear sight aperture. This only applies to iron sights.
- Place any ejected round into the magazine, return the magazine to the magazine pouch, and close the magazine pouch.

Show Clear

On the command **SHOW CLEAR**, the Service rifle is unloaded with a secondary inspection as follows:

- Ensure that the weapon is on **SAFE**.
- Remove the magazine from the Service rifle and retain it.
- Bring your support hand back against the magazine well so that the slip ring rests in the V of your hand. While cupping your support hand under the ejection port, rotate the Service rifle so that the ejection port is facing the deck.
- Pull the charging handle to the rear and catch the round in your support hand.
- Lock the bolt to the rear.
- Put the weapon on **SAFE** if the selector lever would not move to **SAFE** earlier.

- Ensure that the chamber is empty and no ammunition is present.
- Have a second party observer inspect the weapon to ensure that no ammunition is present (see fig. 4-6). The second party observer—
  - –Inspects the chamber visually to ensure that it is empty (i.e., no ammunition is present) and that the magazine is removed.
  - –Places his pinky finger into the chamber to ensure that it is empty.
  - –Ensures that the weapon is on **SAFE**.
  - –Acknowledges that the Service rifle is clear. After receiving acknowledgment that the Service rifle is clear, the Marine—
    - o Depresses the bolt catch and observes the bolt moving forward on an empty chamber.
    - o Closes the ejection port cover.
    - o Checks the sights for proper BZO setting and correct rear sight aperture. This only refers to iron sights.
    - o Places any ejected round into the magazine, returns the magazine to the magazine pouch, and closes the magazine pouch.

Figure 4-6. Observer Inspection.

Filling, Stowing, and Withdrawing Magazines

Filling the Magazine

The magazine can be filled using loose rounds and a 10-round stripper clip.

Loose Rounds

Perform the following steps to fill the magazine with loose rounds:

- Remove a magazine from the magazine pouch.
- Place a round on top of the follower.
- Press down until the round is held between the follower and feed lips of the magazine (see fig. 4-7 on page 4-8).

Figure 4-7. Loose Rounds.

- Repeat until the desired number of rounds is inserted. The recommended number of rounds per magazine is no more than 29. Thirty rounds in the magazine can prohibit the magazine from seating properly on a closed bolt.
- Tap the back of the magazine to ensure the rounds are seated against the back of the magazine.

> *Note:* Ammunition in filled magazines should be rotated frequently if the rounds are not expended. The spring tension on the magazine can weaken if compressed for a long period of time and can cause a malfunction of the magazine. Downloading and cleaning of magazines should be accomplished during preventive maintenance.

10-Round Stripper Clip and Magazine Filler

Perform the following steps to fill the magazine with the 10-round stripper clip and magazine filler (see figs. 4-8 and 4-9):

- Remove a magazine from the magazine pouch.
- Slide the magazine filler into place.
- Place a 10-round stripper clip into the narrow portion of the magazine filler.
- Use thumb pressure on the rear of the top cartridge and press down firmly until all 10 rounds are below the feed lips of the magazine.
- Remove the empty stripper clip while holding the magazine filler in place.
- Repeat until the desired number of rounds is inserted. The recommended number of rounds per magazine is no more than 29. Thirty rounds in the magazine can prohibit the magazine from seating properly on a closed bolt.

- Remove the magazine filler and retain it for future use.
- Tap the back of the magazine to ensure the rounds are seated against the back of the magazine.

Figure 4-8. Magazine Filler and 10 Rounds.

Figure 4-9. Filling the Magazine with a Stripper Clip and Magazine Filler.

Stowing Magazines

Issued Gear

The following applies to the positioning of issued gear:

- When using issued gear with magazines stowed on the chest, filled magazines are stowed with rounds down and projectiles pointing toward either side of the body, depending on the Marine's preference.

- The magazine pouches can be mounted on either side, depending on the Marine's preference and manipulation requirements. However, a majority of magazines should be mounted on the support side to facilitate support-side speed reloads.
- The vest should be tight to the body to prevent shifting, and the magazine pouches should be mounted where they are easily and naturally accessible.

See figure 4-10.

Figure 4-10. Issued Gear.

Magazine Pouch

In a magazine pouch, filled magazines are stored with rounds down and projectiles pointing away from the body. The magazine pouch is worn on a cartridge belt or a load-bearing vest. The belt should be tight around the waist to ensure that the magazine pouch does not drift out of position unexpectedly. Magazine pouches can be placed on either side of the body or both sides, depending on the Marine's preference and manipulation method.

Empty or Partially-Filled Magazines

When empty or partially-filled magazines are stored in a magazine pouch, they are stowed with the rounds or follower up, to allow the selection of filled magazines by touch (e.g., at night, feeling a baseplate indicates a filled magazine). Empty or partially-filled magazines can also be stored in a dump pouch.

Withdrawing Magazines

When withdrawing magazines, the support hand of the Marine is used to withdraw magazines from the magazine pouch, while the firing hand ensures positive control of the Service rifle by maintaining a firm pistol grip.

Withdrawing magazines for an initial or tactical load is considered an administrative function. When conducting a speed reload, time is critical, and the Marine should immediately assume a grip that will facilitate a speed reload. To withdraw magazines from a magazine pouch, the Marine should—

- Pinch the magazine pouch release to open the magazine pouch or unsnap the pouch cover using the thumb and index finger.
- Slide the thumb over the top of the magazine until it rests on the back of the magazine.
- Grasp the magazine with the thumb, little finger, and ring finger and lift the magazine directly out of the pouch.

Grip with all of the fingers and rotate the magazine up to check the number of rounds as the magazine clears the pouch.

Reloading the Service Rifle

Principles of Reloading

When performing a reload, the first priority is to reload the Service rifle and get it back into action. The second priority is to retain the magazine so that when you move, the magazine moves with you. When time permits, retain magazines securely on your person (e.g., cargo pocket, load-bearing vest). The combat situation may dictate dropping the magazine to the deck when performing a speed reload. This is acceptable as long as the magazine is picked up before moving on to another location and if the tactical situation permits.

CAUTION

The dropped magazine should not be dirty or damaged, since these conditions could cause a stoppage.

While reloading—

- The focus should remain on the tactical situation (i.e., keep both eyes and the muzzle focused on the **adversary** while conducting a magazine exchange).
- The weapon should be close to your body while maintaining a firm pistol grip for positive control of the Service rifle. Always use your support hand to make magazine exchanges.
- The new magazine should be tugged on to ensure that it is seated. Do not slam the magazine into the weapon hard enough to cause a round to partially pop out of the magazine. This action will cause a double feed and require corrective action.

When there is a lull in the action—

- Refill the empty magazines for future use.
- Replace any magazines that are low on ammunition. This ensures that there is a full magazine of ammunition in the Service rifle should action resume. Do not wait until the magazine is completely empty to replace it.

Tactical Reload

A tactical reload occurs when the magazine is replaced before it runs out of ammunition, there is a lull in the action, and when the weapon is in Condition 1. To perform a tactical reload—

- Withdraw a filled magazine from the initial load pouch (or the next furthest away magazine pouch) by grasping the filled magazine with an extremely low grip.
- Grasp (high on the magazine with the thumb and fingers) the magazine to be replaced, controlling both the filled and empty magazines with the same hand.
- Depress the magazine's release button to remove the magazine:
  - –Right-handed Marines should press the magazine release with the index finger of the firing hand and remove the magazine with the support hand.
  - –Left-handed Marines should bring the support hand to the slip ring and wrap the hand around the magazine well. Next, press the magazine release button with the thumb of the support hand, and remove the magazine with the support hand.

- Insert the magazine fully into the magazine well until the magazine catch engages the magazine.
- Store the partially-filled magazine in a dump pouch or cargo pocket. Empty or half-empty magazines should never be stored in magazine pouches.

Speed Reload

A speed reload is required when the magazine in the weapon has been emptied and the bolt has locked to the rear. It is conducted as quickly as possible.

- To perform a speed reload, keep the trigger finger straight, press the magazine release button, and remove the empty magazine as follows:
  –Right-handed Marines should press the magazine release with the index finger of the firing hand.
  –Left-handed Marines should press the magazine release with the thumb of the non-firing hand.

> *Note:* While on the move, if the bolt is locked to the rear, it is imperative to continue moving while reloading.

- Insert a filled magazine into the magazine well and tug downward on the magazine to ensure that it is properly seated.
- Depress the bolt catch to allow the bolt carrier to move forward. Observe the round being chambered:
  –Right-handed Marines should strike the upper portion of the bolt catch with the palm of their support hand.
  –Left-handed Marines should slide their nonfiring hand around the front of the magazine well and strike the upper portion of the bolt catch with the fingers of their nonfiring hand. This places the Service rifle in Condition 1.

Corrective Action

If the Service rifle fails to fire, corrective action should be performed by the Marine. Corrective action is the process of investigating the cause of the stoppage, clearing the stoppage, and returning the weapon to proper operating status.

Once the Service rifle ceases firing, the Marine must physically or visually observe the ejection port to identify the problem before he can clear it. The steps taken to clear the weapon are based on observation of one of the indicators discussed in the following subparagraphs:

Indicator: Bolt is Forward or Ejection Port Cover is Closed

The Marine will depress the trigger and hear a click and feel the hammer fall without a shot being fired (see figs. 4-11 and 4-12).

Figure 4-11. Bolt Forward.

Figure 4-12. Ejection Port Cover Closed.

To return the weapon to proper operating status, first seek cover if the tactical situation permits, and then—

- Tap (tap the bottom of the magazine).
- Rack (pull the charging handle to the rear and release it).
- Bang (sight in and attempt to fire).

Indicator: Bolt is Locked to the Rear

Although a dry weapon is not considered a true stoppage or mechanical failure, the Marine must take action to return the weapon to operation. If the Marine observes that the bolt is locked to the rear (see fig. 4-13), the weapon has run dry, and the Marine will perform the following steps to return the weapon to normal operation:

- Conduct a speed reload (see pg. 4-12).
- Sight in and attempt to fire.

Figure 4-13. Bolt Locked to the Rear.

Indicator: Visible Obstruction

A visual obstruction (see fig. 4-14) usually indicates a failure to eject or a double feed. This occurs when a round and a piece of brass become stuck in the chamber or two rounds become stuck in the chamber. The Marine will attempt to depress the trigger and will feel a mushy trigger or the Marine can feel the weapon fail to completely cycle.

To return the weapon to operation—

- Seek cover if the tactical situation permits.
- Pull the charging hand and attempt to lock it to the rear.
- Hold the charging handle to the rear, rotate the Service rifle so that the ejection port is facing down, and shake the Service rifle to free the brass/round. Maintain pressure to keep the charging handle to the rear.

- Attempt to remove the magazine if the brass/round does not shake free. Maintain pressure to keep the charging handle to the rear and hold it. Strike the butt of the Service rifle on the ground or manually clear the brass/round.

> *Note:* The stock on the M4 must be fully collapsed before striking the butt on the deck.

- Conduct a reload.
- Sight in and attempt to fire.

Indicator: Brass is Stuck Above the Bolt

When brass is stuck above the bolt (see fig. 4-15 on page 4-16), the Marine will perform the following steps to clear and return the weapon to operation:

- Seek cover if the tactical situation permits.
- Attempt to place the weapon on **SAFE**.
- Remove the magazine.
- Pull the charging handle to the rear until resistance is met and hold it.
- Rotate the Service rifle so that the ejection port is facing you.
- Push the bolt face back with a sturdy object (e.g., stripper clip, knife, multi-purpose tool).
- Rotate the muzzle down, and observe the brass clearing the chamber.
- Perform a reload.
- Sight in and attempt to fire.

Figure 4-14. Visible Obstruction.

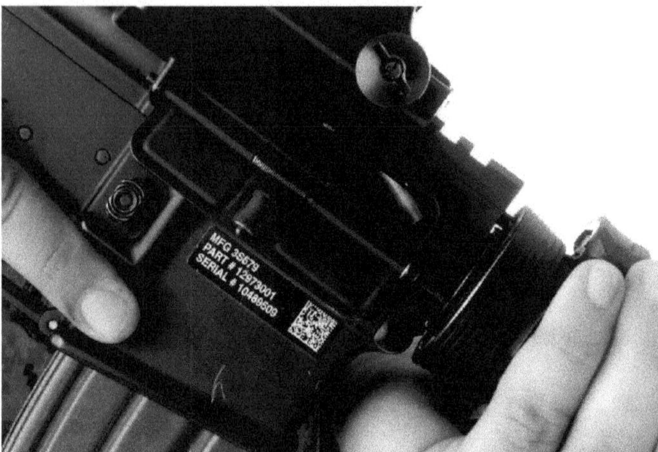

Figure 4-15. Brass Above Bolt.

Indicator: Audible Pop, Reduced Recoil, or Black Smoke

An audible pop occurs when only a portion of the propellant is ignited. It is normally identifiable by reduced recoil and is sometime accompanied by excessive black smoke escaping from the chamber area. To clear the Service rifle in a combat environment—

- Place the Service rifle in Condition 4.
- Move the take-down pin from left to right as far as it will go to allow the lower receiver to pivot.
- Remove the bolt-carrier group.
- Inspect the bore for obstruction from the chamber end.
- Insert a cleaning rod into the bore from the least blocked end to clear the obstruction, pushing in the direction that requires the least amount of travel. This may require striking the cleaning rod with a hard object to push the projectile through the barrel.
- Reassemble the Service rifle.
- Conduct a reload.
- Sight in and attempt to fire.

Weapons Carries

Weapons carries provide an effective way to handle the Service rifle while remaining alert to enemy engagement. Weapons carries are tied to threat conditions and are assumed in response to a specific threat situation. The weapons carry that is assumed prepares the Marine, both mentally and physically, for target engagement. The sling provides additional support for the weapon when firing; therefore, the hasty or three-point sling should be used in conjunction with the carries.

Three-Point Sling Controlled Carry

This carry is used when no immediate danger is present (see fig. 4-16), and the weapon—

- Is on **SAFE**.
- Hangs muzzle down in front of the body. The muzzle should point down, just to the outside of the feet, with the buttstock at approximately armpit level. The Marine maintains constant muzzle awareness.
- Is controlled with the firing hand grasping the pistol grip.

Figure 4-16. Three-Point Cling Controlled Carry.

Tactical Carry with the Web Sling

The Marine carries the Service rifle at the tactical carry if no immediate danger is present. The tactical carry permits control of the Service rifle while the Marine is moving, still allowing quick engagement of the enemy. To assume the tactical carry, the Marine will perform the following steps:

- Place the support hand on the rail system, firing hand around the pistol grip, trigger finger straight along the receiver (see fig. 4-17), and firing hand thumb on top of the selector lever (see fig. 4-18).

Figure 4-17. Straight Trigger Finger.

Figure 4-18. Thumb on Selector Lever.

- Place the buttstock along the side of the body at approximately hip level.
- Angle the muzzle upward approximately 45 degrees, in a safe direction.
- Position the muzzle in front of the eyes, slightly below eye level.

See figure 4-19.

Figure 4-19. Tactical Carry.

Alert Carry with the Web Sling

The Marine should carry the Service rifle at the alert if enemy contact is anticipated. The weapon is on **SAFE** in the alert carry and allows immediate target engagement. The alert carry is used for moving in close terrain (e.g., urban, jungle). To assume the alert, the Marine will—

- Place his support hand on the rail system, his firing hand around the pistol grip, his trigger finger straight along the receiver, and his firing thumb on top of the selector lever.
- Place the buttstock in his shoulder.
- Lower the sights and angle the muzzle downward at 45 to 70 degrees based upon the need for mobility, observation, and muzzle awareness.
- Point the muzzle in a safe direction or the general direction of anticipated enemy contact.

See figure 4-20 on page 4-20.

Ready Carry with the Web Sling

The Marine carries the Service rifle at the ready if enemy contact is imminent. The weapon is on **SAFE** in the ready carry and allows immediate target engagement. To assume the ready, the Marine will perform the following:

- Place his—
  –Support hand on the rail system.
  –Firing hand around the pistol grip.

–Trigger finger straight along the receiver.

–Firing thumb on top of the selector lever.

- Place the buttstock in his shoulder.
- Lower the sights to just below eye level so that a clear field of view is maintained for target identification (see fig. 4-21).
- Point the muzzle in a safe direction or the general direction of imminent enemy contact.

Figure 4-20. The Alert Carry.

Figure 4-21. Ready Carry.

Weapons Transports

Weapons transports are used—

- To carry the Service rifle over the back or shoulders when moving for long periods.
- To provide a more relaxed position for walking.
- If no immediate danger is present.
- If one or both hands are needed for other tasks/work.

Cross Body, Muzzle Up Transport, with Three-Point Sling

This transport carries the weapon across the back with the muzzle up and is employed with the three-point sling. To assume this transport—

- Reach over the top of the rail system and grasp the Service rifle with your support hand.
- Place your firing hand on the buttstock. While pushing down on the buttstock, control and guide the muzzle end of the Service rifle with your support hand, while continuing to push the Service rifle under your support shoulder around to the back.
- Grasp the sling at the shoulder with your firing hand and apply downward tension on the sling. At the same time, continue to guide the buttstock around to the back of the body with your support hand. The Service rifle should be slung across your back with the muzzle up.

See figure 4-22.

Figure 4-22. Cross Body, Muzzle Up Transport with Three-Point Sling.

Strong-Side Sling Arms Transport Muzzle Up with Web Sling

To assume this transport from the tactical carry, perform the following steps:

- Release your hold on the pistol grip.
- Lower the buttstock and bring the Service rifle to a vertical position.
- Grasp the sling above your support forearm with your firing hand.
- Guide the Service rifle around your firing shoulder with your support hand.
- Apply downward pressure on the sling with your firing hand. This stabilizes the Service rifle on your shoulder.
- Release the rail system.

See figure 4-23 on page 4-22.

Weak-Side Sling Arms Transport Muzzle Down with Web Sling

This transport can be used in inclement weather to keep moisture out of the Service rifle's bore. To assume this transport from the tactical carry, perform the following steps:

- Release your hold on the pistol grip.
- Lower the buttstock of the Service rifle, and bring the Service rifle to a vertical position.
- Rotate the Service rifle outboard until the pistol grip is pointing toward your body.
- Reach over the support forearm and grasp the sling with your firing hand
- Rotate the muzzle down with your support hand, while sliding your firing hand up the sling. Place the sling on the support shoulder.
- Grasp the sling with your support hand and apply downward pressure on the sling. This stabilizes the Service rifle on your shoulder.
- Release the rail system.

See figure 4-24.

Cross-Body Sling Arms Transport with Web Sling

The Marine uses this transport if both hands are required for other tasks or work. The Service rifle is slung across the back with the muzzle pointing either up or down.

> *Note:* Normally, the Service rifle is carried with the muzzle down to prevent pointing the muzzle in an unsafe direction.

Figure 4-23. Strong-Side Sling Arms Transport with Muzzle Up.

Figure 4-24. Weak-Side Sling Arms Transport with Muzzle Down.

To assume this transport, perform the following:

- Grasp the sling with your firing hand.
- Grasp the rail system with your support hand.
- Pull up on the Service rifle with both hands.
- Slide the sling over your head.
- Position the Service rifle comfortably across your back.

See figure 4-25.

To assume a cross-body sling transport, perform the following from strong-side sling arms (muzzle up):

- Grasp the sling with your support hand.
- Grasp the pistol grip with your firing hand.
- Pull up on the Service rifle with both hands.
- Slide sling over your head.
- Position the Service rifle so that it rests comfortably across your back.

See figure 4-26.

Figure 4-25. Cross-Body Sling Arms Transport with Muzzle Down.

Figure 4-26. Cross-Body Sling Arms Transport with Muzzle Up.

Carries and Transports with the One-Point Sling

Units are authorized to purchase one-point slings. The unit commanders will determine what the carry procedures are for one-point slings.

Transferring the Service Rifle

Proper weapons handling is required every time the Marine picks up a weapon, passes a weapon to another Marine, or receives a weapon from another Marine. It is the responsibility of the Marine receiving or taking charge of a weapon to determine its condition. Depending on the situation, there are two procedures that can be used to transfer a Service rifle from one Marine to another: show clear transfer and condition unknown transfer.

Show Clear Transfer

When time and the tactical situation permit, the Marine should transfer the Service rifle using the show clear transfer. To properly pass a Service rifle between Marines, the Marine handing off the Service rifle must perform the following:

- Ensure that the Service rifle is on SAFE.
- Remove the magazine if one is present.
- Lock the bolt to the rear.
- Inspect the chamber visually to ensure that there is no ammunition present.
- Leave the bolt locked to the rear and hold it at eye level for the other Marine to visually inspect the bore before handing the weapon to the other Marine, stock first (see fig. 4-27).

The Marine receiving the weapon must—

- Ensure that the Service rifle is on SAFE.
- Inspect the chamber visually to ensure that there is no ammunition present.
- Release the bolt catch and observe the bolt going forward on an empty chamber.
- Close the ejection port cover.

Figure 4-27. Visual Inspection During Show Clear.

Condition Unknown Transfer

There are times when time or the tactical situation does not permit a show clear transfer of the Service rifle. The procedures for the condition, known as unknown transfer, are conducted by the Marine who is taking charge of a Service rifle in a situation when the condition of the Service rifle is unknown (e.g., an unmanned Service rifle from a casualty, a Service rifle stored in a rifle rack).

Note: This transfer can also be conducted when the Marine intentionally transfers a Condition 1 weapon to another Marine. The same considerations for receiving the weapon apply.

To properly take charge of a Service rifle when its condition is unknown, the Marine must perform the following procedures:

- Ensure that the Service rifle is on **SAFE**.
- Conduct a chamber check to determine or verify the condition of the weapon (see pgs. 4-2 and 4-3).
- Remove the magazine and observe or verify that ammunition is present. If time permits, count the rounds.
- Insert the magazine into the magazine well and close the ejection port cover.

Clearing Barrel Procedures

The sole purpose of clearing a barrel is to provide a safe direction in which to point a weapon in a controlled environment when loading; unloading; and unloading, showing clear (see fig. 4-28). Clearing barrel procedures are identical to the weapons handling procedures for the Service rifle when performing load; make ready; unload; unload, show clear.

Figure 4-28. Clearing the Barrel.

CHAPTER 5

Fundamentals of Marksmanship

The fundamentals of marksmanship are aiming and trigger control. These techniques provide the foundation for all marksmanship principles and skills. For rifle fire to be effective, it must be accurate. A rifleman who merely sprays shots in the vicinity of the enemy produces little effect. The fundamentals of precision marksmanship, when applied correctly, form the basis for delivering accurate fire on enemy threats. These skills must be developed so that they are applied instinctively.

During combat, the fundamentals of marksmanship must be applied in a timeframe consistent with both the size and distance of the adversary. At longer ranges, the adversary appears to be smaller and a more precise shot is required to accurately engage the adversary. As the range to the target increases, the fundamentals are more critical to accurate engagement. To be accurate at longer ranges, the Marine must take the time to slow down and accurately apply the fundamentals. At shorter ranges, the enemy must be engaged quickly before he can engage the Marine. As the size of the target increases and the distance to the target decreases, the fundamentals, while still necessary, become less critical to accuracy.

Aiming with Rifle Combat Optic

The Service rifle is defined as a M16A4 rifle or M4 carbine equipped with an RCO. The primary sight for the Service rifle is the RCO. The Service rifle is also equipped with a BUIS.

Sight Picture with Rifle Combat Optic

Sight picture is the placement of the optic reticle pattern in relation to the target. The RCO is calibrated to accommodate bullet drop. The reticle pattern of the RCO is a bullet drop compensator with designated aiming points to compensate for the trajectory of the 5.56-mm round at ranges of 100 to 800 meters. This feature eliminates the need for mechanical elevation adjustments on the Service rifle. Sight picture changes are based on the range to the target. To compensate for range to the target, the aiming points seen in figure 5-1, on page 5-2, are used with the RCO. The horizontal mil scale is removed to emphasize the bullet drop compensator by performing the following:

- Hold the tip of the chevron center mass on a target at 100 meters or less.
- Hold the crotch of the chevron center mass on a target at 200 meters.

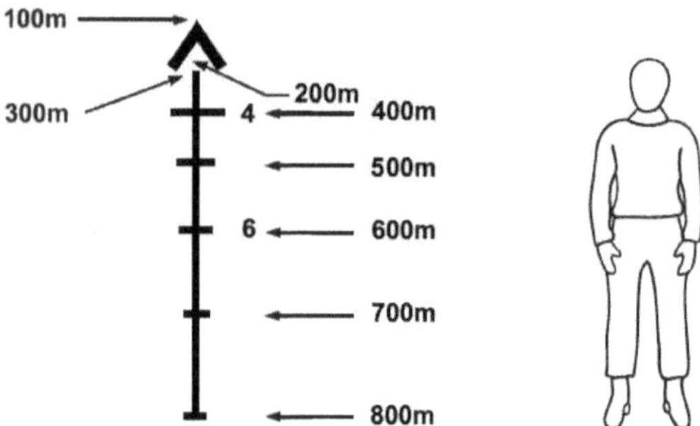

Figure 5-1. Rifle Combat Optic Aiming Points.

- Hold the tip of the red post center mass on a target at 300 meters.
- Hold a horizontal stadia line center mass on a target at each of the ranges indicated beyond 300 meters.

Target Acquisition with Rifle Combat Optic

The RCO is designed for shooting with both eyes open for quick target acquisition and engagement. This allows the RCO to be used as a reflexive sight when speed is critical.

Shooting with Both Eyes Open

Human vision is based upon a binocular presentation of visual information to the brain. This means that the brain processes what is seen through both eyes. The RCO, with its illuminated aiming point and magnification, is designed to present a binocular view of the target; therefore, the RCO is designed to shoot with both eyes open. A traditional scope presents a monocular view. That is why one eye is closed to shoot.

With both eyes open, when the weapon is moved, the brain picks up the chevron in the dominant eye through the optic, and picks up the target and background with the nondominant eye.

During dynamic movement, the scene through the telescope blurs, because the image moves more rapidly due to magnification. The dominant eye sees the bright chevron against the blurred target scene, so the brain picks the scene from the unaided eye and merges the two images. As soon as the weapon begins to become steady in the target area, the brain switches to the magnified view of the target.

Procedures for Determining Dominant Eye

To use the RCO to its maximum potential, the Marine should shoot with both eyes open and use his dominate eye to look through the optic. To determine dominant eye, the following steps should be performed (see fig. 5-2):

Figure 5-2. Determining Dominant Eye.

- Have another Marine stand in front of you approximately 5 to 7 feet away. With both eyes open, look at the Marine and extend your hands at eye level, out to the sides of your body.
- While keeping the hands extended, slowly bring your two hands together, forming a small triangle out in front of your face.
- To determine your dominant eye, the Marine standing in front of you will indicate which of your eyes he sees in the small triangle. For example, if he sees your right eye in the small triangle, you are right-eye dominant. You can confirm this by closing your left eye. You should be able to see the Marine through the triangle with your right eye.

Shooting Adjustment

Marines who are cross-eye dominant (i.e., they use their nondominant eye behind the optic), will experience a shift in point of impact when shooting with both eyes open. The amount of shift will be based on the disparity between the dominant and nondominant eye.

> ***Note:*** If you cannot shoot using the dominant eye behind the optic, keep your dominant eye closed. The downside to this approach is a loss of peripheral vision.

Size and Distance to the Target. During combat, the fundamentals of marksmanship must be applied in a timeframe that is consistent with the size and distance to the target.

Close-Range Engagements. Close-range engagements are those with little or no warning that require an immediate response to engage the enemy. This type of engagement is probable in close terrain (e.g., urban, jungle) and is typically considered to be 50 meters or closer. However, the range is directly dependent upon a Marine's ability: One Marine may characterize close range as being 25 meters because of his abilities; while a more skilled marksman may consider close range as being 50 meters.

At close ranges, when employing the RCO, the Marine should shoot with both eyes open. At close ranges (e.g., 0 to 100 yards)—

- The natural tendency is to look at the target and shoot with both eyes open. Situational awareness is increased by keeping both eyes open.
- The target will appear very large in the magnification of the scope.

- The perfect sight picture is not as critical for accurate target engagement. The weapon is presented rapidly and the shot is fired with the red chevron roughly on the desired target area.
- The enemy must be engaged quickly before he engages the Marine. Marksmanship skills include quick presentation and application of the fundamentals (i.e., quick acquisition of the chevron). To acquire a target, keep both eyes open, focus on the target, and bring the weapon and/or optic up into the line of sight. Do not switch the focus to the reticle. Place the tip of the chevron on the target to achieve sight picture.

Long-Range Engagements. As the range to the target increases, at some point the Marine may have to shoot with one eye closed to achieve the proper sight picture with the RCO reticle pattern. The distance for transitioning from shooting with both eyes open to one eye closed is dependent upon the individual Marine's ability to acquire a precise sight picture. With the nondominant eye closed, the Marine focuses on placing the reticle pattern's aiming point for that range on the target.

At longer ranges, accurate sight picture is more critical to ensure accurate shooting. To be accurate at longer ranges, the Marine must take the time to slow down and accurately apply the fundamentals.

For example, at 300 yards, the tip of the red post in the RCO is held center mass on a target. At this range, it may be difficult to acquire the target, because the red chevron tends to mask the target. Also, there is a tendency to shoot lower at 300 yards, because there is a smaller portion of the target visible. Therefore, ensure that the red post is placed correctly on the target when shooting at this distance. When past 300 yards, the horizontal stadia lines are used as the aiming point. These lines are black and not illuminated, so they must be focused on to establish the sight picture.

Factors Affecting Rifle Combat Optic Sight Picture

Stock Weld

Stock weld is the point of firm contact between the cheek and the stock of the Service rifle (see fig. 5-3).

Figure 5-3. Proper Stock Weld.

The Marine's head should be as erect as possible to enable his aiming eye to look straight through the optic. If the position of his head causes him to look across the bridge of his nose or out from under his eyebrow, his eye will be strained.

> ***Note:*** The eye functions best in its natural forward position. Changing the placement of the cheek either up or down on the stock from shot to shot can affect the zero on the Service rifle because of the perception of the reticle pattern in the optic.

A consistent and proper stock weld is critical to the aiming process, because it provides consistency in eye relief, which affects the ability to obtain sight picture. Consistent placement of the buttstock in the shoulder will assist in achieving a consistent stock weld.

With the RCO, if the butt of the Service rifle is placed in the shoulder correctly and the stock weld is correct, the Marine should be looking through the optic as the Service rifle is presented. As the Service rifle levels, the Marine should see the chevron and establish a sight picture.

Stock weld changes with the Marine's position. The Marine needs to be aware of the ghost image of the front sight post through the optic and keep it as close to the center of the optic as possible in each position. For example, a right-handed Marine's stock weld will be centered on the Service rifle in the prone position and the front sight post will be centered in the lower part of the optic. As the Marine moves to a higher position, the stock weld will move to the left side of the Service rifle and the ghost image of the front sight post will move to the right. Keeping the ghost image of the front sight post as centered as possible will provide a consistency in both zero and shot placement.

Eye Relief

Eye relief is the distance between the optic and the aiming eye (see fig 5-4). Optimal eye relief with the RCO is normally 1 1/2 inches from the optic. The distance between the aiming eye and the optic depends on the size of the Marine and the firing position. While eye relief varies slightly from one position to another, it is important to have the same eye relief for all shots fired from a particular position.

Figure 5-4. Proper Eye Relief with the Rifle Combat Optic.

With the RCO, improper eye relief can cause scope shadow, which can result in improper shot placement, because what appears in the center of the optic is offset by the shadow (see figs. 5-5 and 5-6). With the RCO, if eye relief is too far, scope shadow may occur and the field of view will be smaller, affecting zero and shot placement.

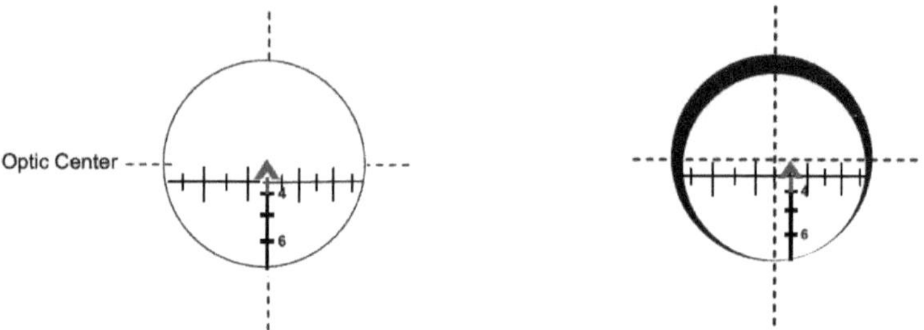

Figure 5-5. No Scope Shadow.

Figure 5-6. Scope Shadow.

Eye relief with the RCO is shorter than with iron sights to open the field of view and eliminate scope shadow. However, eye relief that is too short can cause the RCO to hit the Marine above the eye during recoil. To eliminate scope shadow, adjust eye relief. To adjust the RCO for proper eye relief, use the buddy system and perform the following actions:

- Assume the standing position.
- Have another Marine move the RCO forward and backward on the rail until the Marine finds the optimum position for field of view and no scope shadow. Secure the RCO on the rail.

Glasses, Protective Eyewear, or Goggles

Wearing glasses, ballistic protective eyewear, or sun/wind/dust goggles can alter the perception of sight alignment and sight picture. If wearing glasses, it is critical to look through the optic center of the lens. If wearing eye protection, ensure that the portion being looked through is clear and not scratched.

Aiming with Iron Sights

Sight Alignment with Iron Sights

Sight alignment is a term used when firing iron sights. It is the relationship between the front sight post and rear sight aperture and the aiming eye. See figure 5-7. This relationship is critical to aiming and must remain consistent from shot to shot. To achieve correct sight alignment—

- Center the tip of the front sight post both vertically and horizontally in the rear sight aperture.
- Imagine a horizontal line drawn through the center of the rear sight aperture. The top of the front sight post will appear to touch this line.
- Imagine a vertical line drawn through the center of the rear sight aperture. The line will appear to bisect the front sight post.

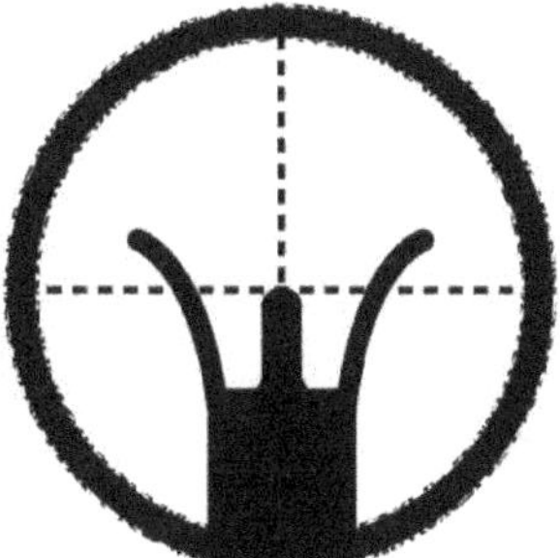

Figure 5-7. Correct Sight Alignment with Iron Sights.

Sight Picture with Iron Sights

Sight picture is the placement of the tip of the front sight post in relation to the target, while maintaining sight alignment. Correct sight alignment, but improper sight placement on the target, will cause the bullet to impact the target incorrectly on the spot where the sights were aimed when the bullet exited the muzzle.

To achieve correct sight picture, place the tip of the front sight post at the center of the target, while maintaining sight alignment (see fig. 5-8). Center mass is the correct aiming point to achieve point of aim and/or point of impact.

Figure 5-8. Correct Sight Picture with Iron Sights.

The sighting systems for the Service rifle are designed to work using a center mass sight picture (see fig. 5-9 on page 5-8).

In combat—

- Targets are often indistinct and oddly shaped. The center mass hold provides a consistent aiming point.
- Targets are often partially exposed as they present themselves from behind cover. A center mass sight picture should still be applied on the largest portion of the target that can be seen (see fig. 5-10 on 5-8).

Figure 5-9. Examples of Correct Sight Picture with Iron Sights.

Figure 5-10. Correct Sight Picture on Partially Exposed Target with Iron Sights.

Correct Sight Alignment Importance

A sight alignment error results in a misplaced shot. The error grows greater proportionately as the distance to the target increases. An error in sight picture; however, remains constant, regardless of the distance to the target (see fig. 5-11).

Sight Alignment and Sight Picture with Iron Sights (Acquiring and Maintaining)

The human eye can only focus clearly on one object at a time. To ensure accurate shooting, observe the following:

- *Focus*. When the shot is fired it is important to focus on the tip of the front sight post.
- *Peripheral vision*. Peripheral vision includes the rear sight and the target. The rear sight and the target will appear blurry. Staring or fixing the vision on the front sight post for longer than a few seconds can distort the image, making it difficult to detect minute errors in sight alignment.
- *Stock weld and Service rifle butt placement*. Proper stock weld and placement of the Service rifle butt in the shoulder aids in establishing sight alignment quickly. The placement of the Service rifle butt in the shoulder serves as the pivot point for presenting the Service rifle up to a fixed point on the cheek (i.e., stock weld).

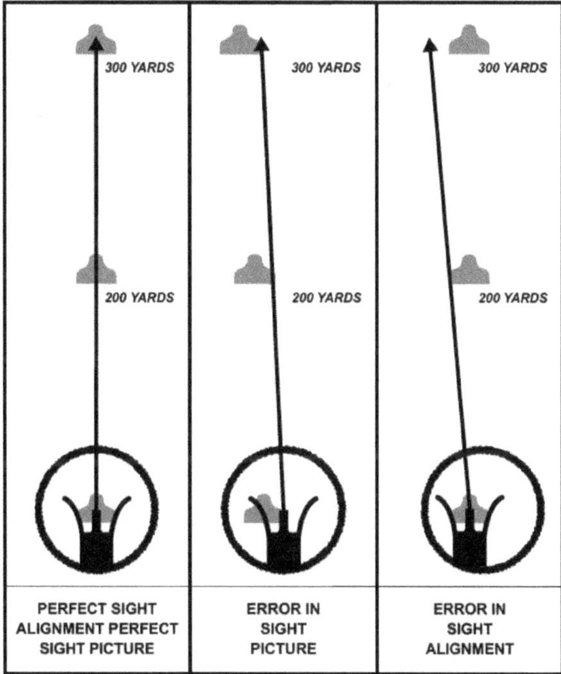

Figure 5-11. Importance of Correct Sight Alignment.

During combat—

- As the Service rifle is presented, the Marine will look at the target. As the Service rifle sights become level with the aiming eye, the Marine will visually locate the target through the rear sight aperture.
- As the Service rifle settles, the Marine's focus shifts back to the front sight post to place the tip of the post on the target and obtain sight alignment and sight picture.
- To maintain sight alignment and sight picture, the Marine's focus should shift repeatedly from the front sight post to the target until correct sight alignment and sight picture are obtained. This will enable the detection of minute errors in sight alignment and sight picture.

To acquire and maintain sight alignment and maintenance, size and distance to the target and close-range engagements should also be employed.

Size and Distance to the Target

During combat, the fundamentals of marksmanship must be applied in a timeframe consistent with the size and distance to the target.

Close-Range Engagements

Close-range engagements are those with little or no warning that require an immediate response to engage the enemy. This type of engagement is probable in close terrain (e.g., urban, jungle) and is typically 50 meters or closer. However, the range is directly dependent upon a Marine's ability. For example, one Marine may characterize close range as 25 meters because of his abilities, while a more skilled marksman may consider close range to be 50 meters.

If this type of engagement is anticipated, the large rear sight aperture (0-2) should be used to provide a wider field of view and detection of targets. Marksmanship skills include quick presentation and application of the fundamentals (i.e., quick acquisition of sight picture).

At close ranges, perfect sight alignment is not as critical to the accurate engagement of targets. The weapon is presented rapidly and the shot is fired with the front sight post placed roughly center mass on the desired target area. The front sight post must be in the rear sight aperture. Proper sight alignment is always the goal.

At close ranges, the enemy must be engaged quickly before he engages the Marine. As distance to the target decreases, the size of the target appears to increase and sight alignment becomes less critical to accuracy.

At very short ranges, a deviation in sight alignment can still produce accurate results as long as the tip of the front sight post is in the rear sight aperture and on the target (see fig. 5-12). The time required to engage a target is unique to each individual. Although the Marine must engage the target rapidly, some semblance of sight alignment is still required to be accurate.

Figure 5-12. Sight Picture at Close-Range Engagements.

Long-Range Engagements

At longer ranges, the target appears smaller and a more precise shot is required to accurately engage the target. Sight alignment and sight picture are more critical to accurate engagement as the range to the target increases. To be accurate at longer ranges, the Marine must—

Figure 5-13. Sight Picture at Long-Range Engagements.

- Take the time to slow down and accurately apply the fundamentals.
- Align the sights with a focus on the tip of the front sight post. Sight picture must be placed correctly on the target.

As the distance to the target increases, the front sight post covers more of the target, making it difficult to establish a center of mass hold (see fig. 5-13). Since the Marine must see the target to engage it, there is a tendency to look at the target by lowering the tip of the front sight post. This causes shots to impact low or miss the target completely. Therefore, the Marine must consciously aim at center of mass and attempt to maintain a center mass sight picture.

Factors Affecting Sight Alignment and Sight Picture with Iron Sights

The following subparagraphs affect sight alignment and sight picture when using iron sights.

Stock Weld

Stock weld is the point of firm contact between the cheek and the stock of the Service rifle (see fig. 5-14 on page 5-12). The head should be as erect as possible to enable the aiming eye to look straight through the rear sight aperture. If the position of the head causes the Marine to look across the bridge of his nose or out from under his eyebrow, the aiming eye will be strained.

> ***Note:*** The eye functions best in its natural forward position. Changing the placement of the cheek up or down on the stock from shot to shot can affect the zero on the Service rifle because of the perception of the rear sight aperture. A consistent and proper stock weld is critical to the aiming process because it provides consistency in eye relief, which affects the ability to align the sights. Consistent placement of the buttstock in the shoulder will assist in achieving a consistent stock weld.

Eye Relief

Eye relief is the distance between the rear sight aperture and the aiming eye (see fig. 5-15 on page 5-12). Normal eye relief is 2 to 6 inches from the rear sight aperture. The distance between the aiming eye and the rear sight aperture depends on the size of the Marine and the firing position. While eye relief varies slightly from one position to another, it is important to have the same eye relief for all shots fired from a particular position.

Figure 5-14. Proper Stock Weld.

Figure 5-15. Proper Eye Relief with Iron Sights.

If the aiming eye is too close to the rear sight aperture, it will be difficult to line up the front sight post in the rear sight aperture (see fig. 5-16). Moving the aiming eye back from the rear sight aperture will make the aperture appear smaller and allow the tip of the front sight post to be easily lined up inside the rear sight aperture.

If the aiming eye is too far from the rear sight aperture, it will be difficult to acquire the target and maintain a precise aiming point (see fig. 5-17).

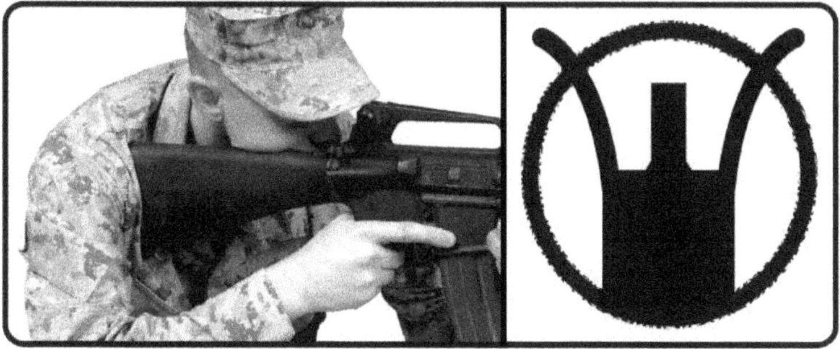

Figure 5-16. Shortened Eye Relief.

Figure 5-17. Extended Eye Relief.

Glasses, Protective Eyewear, and Goggles

Wearing glasses, ballistic protective eyewear, and sun/wind/dust goggles can alter the perception of sight alignment and sight picture. If wearing glasses, it is critical to look through the optic center of the lens. If wearing eye protection, ensure that the portion of the lens that you are looking through is clear and not scratched.

Trigger Control

Trigger control is the skillful manipulation of the trigger that causes the Service rifle to fire without disturbing sight alignment or sight picture. Trigger control is a reaction to what is seen through the sight(s). Controlling the trigger is a mental process, while moving the trigger is a physical process.

Grip

A firm grip is essential for effective trigger control. The grip is established before starting the application of trigger control and it is maintained throughout the duration of the shot. To establish a firm grip on the Service rifle, position the V formed between the thumb and index finger on the pistol grip, behind the trigger. The fingers and thumb are placed around the pistol grip in a location that allows the trigger finger to be placed naturally on the trigger and the thumb in a position to operate the safety (see fig. 5-18). Once established, the grip should be firm enough to allow manipulation of the trigger straight to the rear, without disturbing the sights.

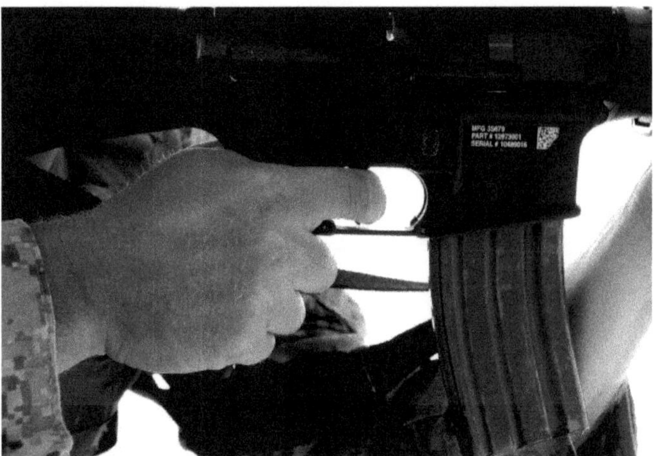

Figure 5-18. Grip.

Trigger Finger Placement

Correct trigger finger placement allows the trigger to be pulled straight to the rear without disturbing sight alignment. The trigger finger should contact the trigger naturally (see fig. 5-19), but not the Service rifle receiver or trigger guard.

Types of Trigger Control

There are two techniques of trigger control: uninterrupted and interrupted.

Uninterrupted Trigger Control

The preferred method of trigger control is uninterrupted. After obtaining sight picture, the Marine applies smooth, continuous pressure rearward on the trigger until the shot is fired.

Interrupted Trigger Control

Interrupted trigger control is used at any time the sight alignment is interrupted or the target is temporarily obscured. An example of this is when extremely windy conditions will not allow the

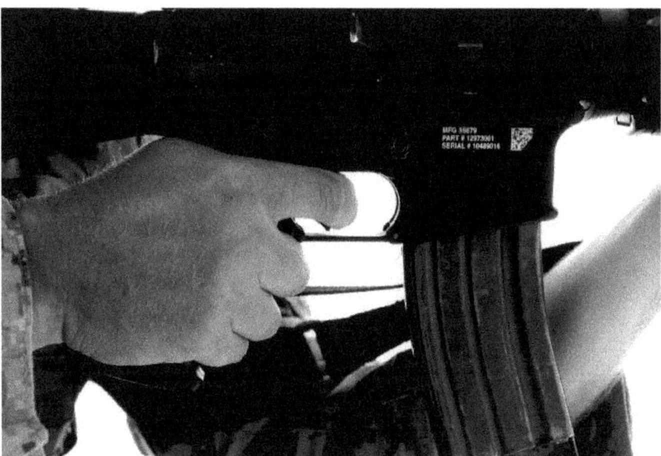

Figure 5-19. Trigger Finger Placement.

weapon to settle, forcing the Marine to pause until the sights return to the aiming point. To perform interrupted trigger control—

- Move the trigger to the rear until an error is detected in the aiming process. Stop the rearward motion on the trigger when this occurs, but maintain the pressure on the trigger until sight picture is achieved.
- Ensure that the sight picture has settled, then continue the rearward motion on the trigger until the shot is fired.

Resetting the Trigger

During recovery, release the pressure on the trigger slightly to reset the trigger after the first shot is delivered. This will be indicated by an audible click. Do not remove the finger from the trigger. This promotes consistency in trigger control and places the trigger in position to fire the next shot without having to reestablish trigger finger placement.

Close-Range Engagements

At close ranges, trigger control should be applied quickly and instantly the moment that sight picture is achieved. Trigger control is applied as sight picture is being acquired.

Mid- to Long-Range Engagements

In mid- to long-range engagements, trigger control is more critical to ensure accuracy. At longer ranges, if the trigger is moved so that sight picture is disturbed, there is a greater chance of missing the target. The longer the range, the more amplified any error in sight picture becomes. The trigger must be manipulated directly to the rear without disturbing sight picture.

CHAPTER 6

Service Rifle Firing Positions

In a combat environment, the Marine must be prepared to engage the enemy under any circumstance. There are four basic firing positions: prone, sitting, kneeling, and standing. These positions provide a stable foundation for effective shooting. Any firing position must provide stability, mobility, and observation of the enemy. During training, the Marine learns positions in a step-by-step process, guided by a series of precise movements until the Marine assumes a correct position. The purpose of this process is to ensure that the Marine correctly applies all of the factors that will assist him in holding the Service rifle steady. The Marine will gradually become accustomed to the feel of the positions through practice and eventually will know instinctively if his position is correct. In combat, it may not be possible to assume a textbook firing position because of terrain, available cover, engagement time, dispersion of targets, and other limiting factors. Modifications to the basic positions may need to be made to adjust to the combat environment. The Marine must strive to assume a position that offers stability for firing, maximum cover and concealment from the enemy, and maximum observation of the target.

Stability of Hold

A firing position must provide a stable platform for accurate and consistent shooting. If the position is solid, the sight(s) can be held steady so sight picture can be achieved and the target accurately engaged. The purpose of a good firing position is to achieve stability of hold. The ability to hold the Service rifle sight(s) still on a designated area of a target is considered stability of hold. Stability of hold is much more apparent when firing the RCO because the 4-power scope magnifies the movement of the reticle pattern on the target. Movement of the sight(s) is not detected as easily with iron sights as it is with the RCO. When using the RCO to engage targets at longer distances, the magnification shows movement of the sight more because of stability of hold, which can slow down reaction time. Firing positions, sling adjustment, and the use of support affect stability of hold and the ability to achieve it. The firing position must be stable enough to hold the Service rifle sight(s) on either a point or an area that is located on the target. Size and distance to the target dictate how critical stability of hold must be.

Smaller Target/Longer Range

If the target is smaller and/or the range is longer, the target will require more stability of hold. As the range to the target increases, the appearance of the target becomes smaller, making stability of hold much more critical to accurate target engagement. The greater the stability of hold, the less the movement of the sight(s) on the target. At longer ranges, the area on the target where the sight(s) are placed is smaller, because the target is further away. A more refined stability of hold is required to keep the sights from moving off of the target.

The shooting process can be slowed down based on the time and distance to the target. As the range to the target increases, there is more time to engage the target and allow stability of hold to be refined for an accurate shot on target. At longer ranges, positions are established to gain stability of hold through the use of bone and artificial support. As time permits, a more stable position (i.e., kneeling or prone) is acquired, and the acquired position should be refined to increase stability of hold. A stable position is critical at longer ranges to acquire the stability of hold that is required to stabilize the sights on the target.

Larger Target/Closer Range

The larger the target and/or the shorter the range, requires less stability of hold; however, the sight(s) must still be stabilized on the target. Stability of hold is not as critical in a close-range engagement as it is for a long-range engagement. Stability of hold allows placement of the Service rifle sight(s) on a target. The greater the stability of hold, the less the movement of the sight(s) on the target. At close ranges, the area on the target where the sight(s) are placed is larger, because the target is closer. The sight's movement area on the target is acceptable as long as the sights do not move off the target. At close ranges there is room for greater movement of the sights on the target; therefore, not requiring as much stability of hold to be accurate. This allows for the quicker presentation and engagement that is required of a close-range target.

Selecting a Firing Position

The selection of a firing position (i.e., prone, sitting, kneeling, and standing) is based on terrain, available cover, dispersion of targets, and other limiting factors. The Marine must select a position that offers stability, mobility, and observation.

Stability

A firing position must provide a stable base for shooting. The prone position is considered the most stable because of the amount of contact with the ground; while the standing position is considered the least stable. A position is selected based on the need for stability of hold. For example, if a precision shot is required, the most stable position should be assumed, as permitted by time and terrain.

At close ranges, positions are established for quick engagement, rather than establishing a perfect position that provides a high degree of stability of hold. Time is of the essence in a close-range engagement, with the intent being to establish rounds on target as quickly as possible. Due to the nature of close engagements, the Marine often fires from a standing position. The Marine assumes an aggressive, mobile standing position with hips, feet, and torso squared to the target; while leaning forward to control recoil of the Service rifle. As time permits, a more stable position can be acquired or the use of support can be incorporated into the position to increase stability.

Mobility

A firing position must provide the Marine with the mobility required to move to new cover or another area. The standing position permits maximum mobility and allows the most lateral movement for engagement of widely dispersed targets. The prone position allows the least mobility and limited lateral movement.

Observation of the Enemy

A firing position must limit the Marine's exposure to the enemy, yet allow observation of the enemy. Manmade structures and terrain features (e.g., vegetation, earth contours) often dictate the firing position.

The standing position normally provides the best field of view, but may allow the most exposure to the enemy. The prone position normally allows the least exposure, but usually provides a limited field of view.

Factors Common to All Firing Positions

There are seven factors common to all firing positions that affect the ability to hold the Service rifle steady, maintain sight alignment and sight picture, and control the trigger. How these factors are applied differs slightly for each position, but the principles of each factor remain the same.

Forward Hand

The placement of the forward hand affects how much muscular tension must be applied to hold the weapon up, affecting stability of hold.

The forward hand should be placed to provide vertical bone support, with the elbow placed under the weapon. The body's skeletal structure provides a stable foundation to support the Service rifle's weight. A weak firing position will not withstand a Service rifle's repeated recoil when firing at the sustained rate or buffeting from wind. To attain a correct firing position, the body's bones must support as much of the Service rifle's weight as possible. Proper use of a sling provides additional support. The weight of the weapon should be supported by bone, rather than muscle, because muscles fatigue, whereas bones do not. Therefore, a balance must be struck between bone support and muscular tension to stabilize the sight. In combat, rearward pressure may be applied slightly, using the forward hand to assist in stabilizing the sight.

The forward hand may grasp the rail system (see fig. 6-1 on page 6-4), the fore grip, or the grip pod. The further out the forward hand is placed on the rail system—

- The less bone support can be achieved, because the elbow cannot be inverted directly under the Service rifle.
- The more muscular tension that is required to hold the weapon up.

Figure 6-1. Forward Hand on Rail System.

If using a fore grip on the weapon, its placement must facilitate stability of hold. The fore grip should be placed on the rail from the center, back toward the receiver, for maximum control and stability of the weapon (see fig. 6-2). The forward hand should pull back on the fore grip, because the rearward pressure helps keep the Service rifle butt in the shoulder.

Figure 6-2. Forward Hand on Fore Grip.

If using a grip pod on the weapon, its placement must facilitate stability of hold. The grip pod should be placed on the rail from the center, back toward the receiver, for maximum control and stability of the weapon (see fig. 6-3). The forward hand should pull back on the grip pod, because the rearward pressure helps keep the Service rifle butt in the shoulder. If the bipod legs are deployed on the grip pod, the forward hand should pull back and down on the grip pod to keep the legs firmly on the deck and/or surface.

Figure 6-3. Grip Pod.

Service Rifle Butt in the Shoulder

Placement of the Service rifle butt firmly in the pocket of the shoulder provides resistance to recoil, helps steady the Service rifle, and prevents the Service rifle butt from slipping during firing (see fig. 6-4). The position must offer a certain amount of resistance in order to allow the weapon to function as designed. If there is not enough resistance to recoil, the weapon may not operate correctly to feed and chamber the next round.

Figure 6-4. Service Rifle Butt in Shoulder.

With the body armor donned, it may be difficult to place the Service rifle butt in the pocket of the shoulder; therefore, the buttstock should be moved to accommodate the jacket. The placement of the buttstock—

- Inboard from the shoulder extends eye relief, making acquisition of the sights more difficult. In addition, eye relief may be more extended, because the body is more squared to the target.
- Outboard from the shoulder reduces eye relief. The Service rifle butt should be placed where it will not slip during recoil, but still allow acquisition of the sights.

> ***Note:*** Some body armors are equipped with a stop that is designed to prevent the Service rifle from slipping during recoil.

To absorb recoil during rapid, multiple engagements in combat, the bulk of the Service rifle butt must be placed in the shoulder. The stock may be lower in the shoulder, affecting stock weld, which also affects eye relief.

Grip of the Firing Hand

A firm grip must be established on the pistol grip to enable rearward pressure to be applied to keep the Service rifle butt in the shoulder (see fig. 6-5). With the body armor donned, additional rearward pressure from the grip may be required to help keep the Service rifle butt firmly in the shoulder, assisting in managing recoil.

Firing-Side Elbow

The firing elbow should be positioned naturally to provide balance to the position and create a pocket in the shoulder for the Service rifle butt. Muscular tension may be increased in the firing arm to hold the Service rifle butt in the shoulder due to the rearward pressure from the grip. The placement of the elbow should remain consistent from shot to shot, ensuring that the resistance to recoil remains constant (see fig. 6-5).

Figure 6-5. Grip and Elbow.

Stock Weld

The placement of the Marine's cheek against the stock should remain firm and consistent from shot to shot. Stock weld is correct when the cheek bone rests on the stock (see fig. 6-6). Consistency of stock weld is achieved in combat through consistent presentation. Presentation to the target stops at stock weld with a firm contact between the cheek and stock. Sight picture

Figure 6-6. Stock Weld.

should be acquired the instant stock weld is achieved. The more squared the position to the target, the more eye relief can extend, making sight acquisition more difficult. To achieve proper eye relief, the head may be rolled forward to reduce the distance between the aiming eye and the sight; however, the head should be held as erect as possible to allow the aiming eye to see directly through the sight(s).

Breathing

Breathing causes movement of the chest and a corresponding movement in the Service rifle and its sight(s). To minimize this movement and the effect it has on aiming, apply breathing control as follows:

- During slow, precision fire where time is not a factor, the shot should be fired during the natural respiratory pause. A respiratory cycle (i.e., inhaling and exhaling) lasts approximately 4 or 5 seconds. Between respiratory cycles there is a natural pause of 2 to 3 seconds. This is the natural respiratory pause. During the respiratory pause, breathing muscles are relaxed and the Service rifle sights settle at their natural point of aim. The Marine should fire at this point.
- During all other engagements (i.e., other than precision fire), take a deep breath, filling the lungs with oxygen as the weapon is presented. Taking in air assists in bringing the weapon up. Hold your breath and apply pressure to the trigger. Holding your breath allows the torso to remain still while controlled muscular tension is applied to stabilize the weapon.

Muscular Control

There are two components of muscular control: controlled muscular tension and muscular relaxation. The type of muscular control required depends on the type of sling, support, and firing position that is being used.

Controlled Muscular Tension: Three-Point Sling

With the three-point sling donned, the Marine must apply an amount of controlled muscular tension in the support arm to stabilize the weapon sight(s). Muscular tension is applied to hold the Service rifle up and steady the sights. Where possible, and as equipment allows, bone support should be achieved to minimize the amount of muscular tension required to hold the weapon. Controlled muscular tension—

- Offers a certain amount of resistance to allow the weapon to function as designed.
- Assists in recoil management.

Following a shot, it is important to concentrate the Service rifle sights back onto the target for another shot. This is known as recovery. Shot recovery begins as soon as the round leaves the barrel. To recover quickly, apply a consistent amount of muscular tension within the position throughout the shot process to allow recovery of the sight(s) back on target as quickly as possible. This allows for rapid reengagement of the enemy.

Muscular Relaxation: Loop Sling

The loop sling is employed during training to facilitate stability of hold so that the fundamentals of marksmanship can be developed and refined. When using the loop sling, the muscles should be relaxed. Relaxation prevents undue muscle strain and reduces excessive movement. If proper relaxation is achieved, natural point of aim and sight alignment/sight picture is easier to maintain.

Elements of a Firing Position When Using a Loop Sling

The loop sling is employed during training to provide maximum stability to enable the fundamentals of marksmanship to be learned and applied. When adjusted properly, the loop sling provides maximum stability for the weapon, and helps hold the front sight/optic reticle pattern still, reducing the effects of the Service rifle's recoil.

Once a sling adjustment is established that provides maximum control of the weapon, the same sling adjustment should be maintained. Varying the sling tension will affect the strike of the bullet, which will make maintaining a BZO difficult. Using the same sling adjustment will ensure the accuracy of rounds on target.

There are three elements of a good firing position that apply when using a loop sling: bone support, muscular relaxation, and natural point of aim.

Bone Support

The body's skeletal structure provides a stable foundation to support the Service rifle's weight. A weak firing position will not withstand a Service rifle's repeated recoil when firing at the sustained rate or buffeting from wind. When possible, the body's bones must support as much of the Service rifle's weight as possible. Proper use of the loop sling provides additional support. The weight of the weapon should be supported by bone rather than muscle, because muscles fatigue; whereas bones do not. An example of this is when the support elbow is resting on the knee in the kneeling position, with the arm directly underneath the weapon, supporting its weight.

By establishing a strong foundation for the Service rifle using bone support, the Marine can relax as much as possible, while still minimizing weapon movement.

Muscular Relaxation

Once bone support is achieved, the muscles become relaxed. Muscular relaxation—

- Helps hold the Service rifle steady and increases the accuracy of the aim.
- Permits the use of maximum bone support to create a minimum arc of movement and consistency in resistance to recoil.

Muscle relaxation cannot be achieved without bone support. During the shooting process, the muscles of the body must be relaxed as much as possible. Muscles that are tense will cause excessive movement of the Service rifle, disturbing the aim. When proper bone support and muscular relaxation are achieved, the Service rifle will settle onto the aiming point, making it possible to apply trigger control, and deliver a well-aimed shot.

Natural Point of Aim

Natural point of aim is the point where the Service rifle sights settle when both bone support and muscular relaxation are achieved. Since the Service rifle becomes an extension of the body, it may be necessary to adjust the position of the body until the Service rifle sights settle naturally on the desired aiming point located on the target. When in a firing position with proper sight alignment, the position of the tip of the front sight post, or the reticle pattern in the RCO, will indicate the natural point of aim.

When completely relaxed, the tip of the front sight post or RCO reticle pattern will indicate the natural point of aim. One method that may be used to check for the natural point of aim is to aim in on the target, close the eyes, take a couple of breaths, and relax as much as possible.

> ***Note:*** When the eyes are opened, the tip of the front sight post/RCO reticle pattern should be positioned on the desired aiming point while maintaining sight alignment.

For each firing position, specific adjustments will cause the Service rifle sights to settle on center mass, achieving a natural point of aim. In all positions, the natural point of aim can be adjusted by—

- Varying the placement of your support hand in relation to the rail system as follows:
  –Moving your support hand forward on the rail system to lower the muzzle of the weapon, causing the sights to settle lower on the target.
  –Moving the support hand back on the rail system to raise the muzzle of the weapon, causing the sights to settle higher on the target.
- Varying the placement of the stock in the shoulder as follows:
  –Moving the stock higher in the shoulder to lower the muzzle of the weapon, causing the sights to settle lower on the target.
  –Moving the stock lower in the shoulder to raise the muzzle of the weapon, causing the sights to settle higher on the target.

- Adjusting the natural point of aim right or left by adjusting body alignment in relation to the target:
  –In the prone position, if the natural point of aim is above or below the desired aiming point, move your body slightly forward or backward (while using your support elbow as a pivot), dig your toes in, push your body forward (causing the sights to settle lower on the target), and pull your body backward (causing the sights to settle higher on the target).
  –In the kneeling and sitting positions, natural point of aim can be adjusted by varying the placement of your support elbow on your knee by moving your support elbow forward on your knee (which lowers the muzzle of the weapon, causing the sights to settle lower on the target) and moving your support elbow back on your knee (which raises the muzzle of the weapon, causing the sights to settle higher on the target).

Firing Positions

Firing positions include prone, sitting, kneeling, and standing.

Prone Position

The prone position provides a very steady foundation for shooting and presents a low profile for maximum concealment. However, the prone position is the least mobile of the firing positions and may restrict the Marine's field of view for observation. In the prone position, the Marine's weight is evenly distributed on his elbows, providing maximum support and good stability for the Service rifle. Depending on the combat situation, the prone position can be assumed by either moving forward or dropping back into position.

Moving Forward into Position

To move forward into the prone position—

- Don the hasty or three-point sling, stand erect, face the target, and place your feet a comfortable distance apart (i.e., approximately shoulder width).
- Place your support hand on the rail system/fore grip/grip pod and your firing hand on the pistol grip.
- Lower yourself into position by dropping to both knees (see fig. 6-7).
- Shift your weight forward to lower the upper body to the ground, using your support hand to break the forward motion (see fig 6-8).

Dropping Back into Position

It may be necessary to drop back into position to avoid either crowding cover or covering unclear terrain. To drop back into the prone position—

- Face the target.
- Place your support hand on the rail system/fore grip/grip pod and your firing hand on the pistol grip.
- Squat to the ground and break your fall with your support hand (see fig. 6-9).
- Kick both legs straight out to the rear (see fig. 6-10).

Figure 6-7.
Dropping to Both Knees.

Figure 6-8.
Moving Forward Into Position.

Figure 6-9.
Breaking the Fall.

Figure 6-10.
Kicking Back Into Position.

Assuming Straight-Leg Prone Position with Three-Point Sling

To assume the straight-leg prone position with the three-point sling—

- Stand erect, face the target, and spread your feet a comfortable distance apart (i.e., approximately shoulder width).
- Grasp the pistol grip with your firing hand.
- Reach under the sling and grasp the rail system/fore grip/grip pod with your support hand in a location that provides maximum bone support and stability for the weapon, but do not incorporate the sling into your grasp.
- Use your firing hand to push the butt of the weapon down to elevate the muzzle.
- Maintain control of the muzzle with your support hand on the rail system/fore grip/grip pod.
- Lower yourself into position by dropping to both knees.
- Release your firing hand from the pistol grip to break the fall, while dropping into position.
- Shift your weight forward to lower your upper body to the ground, using your firing hand to break the forward motion.
- Roll your body to the left, extending and inverting your support elbow on the ground, and stretching your legs out behind.
- Grasp the pistol grip with your firing hand, pulling back on both your firing and support hands to place the Service rifle butt into your firing shoulder, so that the sights are level with your eyes.
- Rotate your body to the right, while lowering your elbow to the ground, so that your shoulders are level.
- Lower your head and place your cheek firmly against the stock in the same place for every shot to ensure consistent eye relief and stock weld.
- Slide your body back to raise the muzzle and push your body forward to lower the muzzle.
- Spread your feet a comfortable distance apart (i.e., approximately shoulder width) with your toes pointing outboard and the inner portion of your feet in contact with the ground. Align as much of your body mass as possible directly behind the Service rifle. If body alignment is correct, the weapon's recoil is absorbed by your whole body and not just your shoulder.

See figure 6-11.

Assuming Straight-Leg Prone Position with Loop Sling

To assume a straight-leg prone position with the loop sling, either move forward or drop back into position as follows:

- Once on the ground, roll your body to the left side as you extend and invert your support elbow on the ground, stretching your legs out behind. Your feet should be spread a comfortable distance apart (i.e., approximately shoulder width) with your toes pointing outboard and the inner portion of your feet in contact with the ground.
- Align as much of your body mass as possible directly behind the Service rifle.
- Recoil of your weapon is absorbed by your entire body and not just your shoulder if body alignment is correct. You should grasp the Service rifle butt with your firing hand and place the Service rifle butt into the shoulder pocket.
- Grasp the pistol grip with your firing hand.

Figure 6-11. Straight-Leg Prone with Three-Point Sling.

- Rotate your body to the right, lowering your elbow to the ground so that your shoulders are level. Your firing hand will pull and hold the Service rifle into your shoulder.
- Lower your head and place your cheek firmly against the stock to allow your aiming eye to look through the rear sight aperture.
- Move your support hand to a location under the rail system, providing maximum bone support and stability for the weapon. This may require that you remove the Service rifle from your shoulder to reposition your support hand.

See figure 6-12 on page 6-14.

Assuming Cocked-Leg Prone Position with Three-Point Sling

To assume the cocked-leg prone position with the three-point sling—

- Drop to the ground in the same manner as for the straight-leg position.
- Once on the ground—
  - –Roll your body to the left, extend and invert your support elbow on the ground.
  - –Stretch your support leg out behind, almost in a straight line. This allows the mass of your body to be placed behind the Service rifle to aid in absorbing recoil.
- Grasp the pistol grip with your firing hand and pull back on both the firing and support hands to place the Service rifle butt into the firing shoulder so that the sights are level with your eyes.
- Turn the toe of your support foot inboard so that the outside of your left leg and foot are in contact with the ground. Next, bend your right leg and draw it up toward your body into a comfortable position. Turn your right leg and foot outboard so that the inside of your right

Figure 6-12. Straight-Leg Prone with Loop Sling.

boot is in contact with the ground. Cocking the leg will raise the diaphragm, making breathing easier.

- Rotate your body to the right, while lowering your elbow to the ground, so that your shoulders are level.
- Lower your head and place your cheek firmly against the stock, in the same place for every shot, to ensure consistent eye relief and stock weld.
- Slide your body back to raise the muzzle and forward to lower the muzzle, pushing your body forward with your feet.

See figure 6-13.

Figure 6-13. Cocked-Leg Prone Position with Three-Point Sling.

Assuming Cocked-Leg Prone Position with the Loop Sling

To assume this position, either move forward or drop back into position and perform the following:

- Once on the ground, roll your body to the left and extend and invert your support elbow on the ground. Your support-side leg should be stretched out behind you, almost in a straight line. This allows the mass of your body to be placed behind the Service rifle to aid in absorbing any recoil.
- Turn the toe of your support-side foot inboard so that the outside of your support-side leg and foot are in contact with the ground. Then bend the firing-side leg and draw it up toward your body into a comfortable position. Turn your firing-side leg and foot outboard so that the inside of the firing-side boot is in contact with the ground. Cocking your leg will raise the diaphragm, making breathing easier.
- Grasp the Service rifle butt with your firing hand and place the Service rifle butt into your shoulder pocket.
- Grasp the pistol grip with your firing hand.

- Roll your body while lowering your firing elbow to the ground. Your firing shoulder should be higher than your support shoulder when in the cocked-leg position.
- Lower your head and place your cheek firmly against the stock to allow your aiming eye to look through the rear sight aperture.
- Move your support hand to a location under the rail system. This will provide maximum bone support and stability for the weapon.

See figure 6-14.

Figure 6-14. Cocked-Leg Prone with the Loop Sling.

Using Support in the Prone Position

If possible, the Marine should use support (i.e., cover and concealment) from a prone position when firing from behind cover. This position is the steadiest and provides the lowest silhouette and maximum protection from enemy fire. Support this position by placing the rail system/fore grip/grip pod, the forearm, or the magazine on or against support (see fig. 6-15).

Figure 6-15. Prone Position Using Support.

The prone position—

- Can be assumed behind almost any type of cover (e.g., tree, wall, log).
- Is flexible and allows shooting from all sides and various sizes of cover.

The body must be adjusted to conform to the cover. For example, if the cover is narrow, keep your legs together. The body should be in line with and directly behind the Service rifle (see fig. 6-16) to present a smaller target to the enemy and provide more body mass to absorb any recoil.

Figure 6-16. Prone Behind Cover.

Sitting Position

There are three variations of the sitting position: crossed-ankle, crossed-leg, and open-leg. The Marine should experiment with all of the sitting position variations and select the position that provides the most stability for firing. Although the sitting position provides a very stable base, it limits lateral movement and maneuverability. The sitting position—

- Provides greater elevation than the prone position, while still having a fairly low profile.
- Has several variations that can be adapted to the individual Marine.

Crossed-Ankle Sitting Position

The crossed-ankle sitting position can be assumed with both the three-point sling and the loop sling.

Three-Point Sling. The crossed-ankle sitting position is a very stable firing position. This position—

- Places most of the body's weight behind the weapon and aids in quick shot recovery.
- Provides a medium base of support and places some of the body's weight behind the weapon for quick recovery after each shot.

To assume the crossed-ankle sitting position with the three-point sling—

- Stand erect and face the target at approximately a 45-degree angle to the target.
- Grasp the pistol grip with your firing hand.
- Grasp the rail system/fore grip/grip pod with your support hand, under the sling in a position that provides maximum bone support and stability of the weapon.

- Bend at your knees and, while elevating the muzzle, drop to one knee and roll back onto your buttocks.
- Extend your legs and cross your support-side ankle over your other ankle.
- Pull back on both your firing and support hands to place the Service rifle butt into your firing shoulder so that the sights are level with your eyes.
- Place your cheek firmly against the stock to obtain a firm stock weld.
- Bend forward at the waist and place your left elbow on your left leg below the knee.
- Lower your right elbow to the inside of your right knee.

See figure 6-17.

Figure 6-17. Crossed-Ankle Sitting Position with Three-Point Sling.

To adjust for elevation—

- Move your support hand rearward or forward on the rail system. Moving your hand rearward elevates the muzzle.
- Move the fore grip/grip pod up in your grasp to raise the muzzle and down in your grasp to lower the muzzle.

To adjust for a minor cant in the Service rifle—

- Rotate the rail system left or right by rotating the pistol grip left or right.
- Rotate the fore grip/grip pod left or right.

Loop Sling. To assume a crossed-ankle sitting position with the loop sling—

- Stand erect and face the target at a 10 to 30 degree angle to the right of the line of fire.
- Place your support hand under the rail system.
- Bend at the knees and break your fall with your firing hand.
- Push backward with your feet to extend your legs and place your buttocks on the ground.
- Cross your support ankle over your other ankle.
- Bend forward at the waist and place your support elbow on your support-side leg below the knee.
- Grasp the Service rifle butt with your firing hand and place the Service rifle butt into your shoulder pocket.

- When the Service rifle butt is seated properly, grasp the pistol grip with your firing hand.
- Lower your firing elbow to the inside of your firing-side knee.
- Place your cheek firmly against the stock to obtain a firm stock weld.
- Move your support hand to a location under the rail system that provides maximum bone support and stability of the weapon.

See figure 6-18.

Figure 6-18. Crossed-Ankle Sitting Position with Loop Sling.

Crossed-Leg Sitting Position

The crossed-leg sitting position can be assumed with both the three-point and loop slings.

Three-Point Sling. To assume the crossed-leg sitting position with the three-point sling—

- Stand erect and face the target at approximately a 45-degree angle to the target.
- Grasp the pistol grip with your firing hand.
- With your support hand, grasp the rail system/fore grip/grip pod under the sling in a position that provides maximum bone support and stability of the weapon.
- Bend at the knees and, while elevating the muzzle, drop to one knee and roll back onto your buttocks.
- Cross your support-side leg over your other leg as close to your buttocks as you can comfortably.
- Pull back on both your firing and support hands to place the Service rifle butt into your firing shoulder so that the sights are level with your eyes.
- Place your cheek firmly against the stock to obtain a firm stock weld.
- Bend forward at the waist while placing your support elbow on your support leg into the bend of the knee.
- Lower your firing elbow to the inside of your firing-side knee.

See figure 6-19.

To adjust for elevation—

- Move your support hand rearward or forward on the rail system. Moving your hand rearward elevates the muzzle.
- Move the fore grip/grip pod up in the grasp to raise the muzzle or move it down in your grasp to lower the muzzle.

Figure 6-19. Crossed-Leg Sitting Position with Three-Point Sling.

To adjust for a minor cant in the Service rifle—

- Rotate the rail system left or right by rotating the pistol grip left or right.
- Rotate the fore grip/grip pod left or right.

Loop Sling. To assume a crossed-leg sitting position with the loop sling—

- Stand erect and face the target at a 45 to 60 degree angle to the right of the line of fire.
- Place your support hand over the sling from the left side and under the rail system.
- Cross your support leg over your other leg.
- Bend at the knees while breaking your fall with your firing hand.
- Place your buttocks on the ground as close to your crossed legs as is comfortable.
- Bend forward at the waist while placing your support elbow on your support-side leg into the bend of the knee.
- Grasp the Service rifle butt with your firing hand and place the Service rifle butt into your shoulder pocket.
- When the Service rifle butt becomes seated, grasp the pistol grip firmly with your firing hand.
- Lower your firing elbow to the inside of your firing-side knee.
- Place your cheek firmly against the stock to obtain a firm stock weld.
- Move your support hand to a location under the rail system that will provide maximum bone support and stability of the weapon.

See figure 6-20.

Figure 6-20. Crossed-Leg Sitting Position with Loop Sling.

Open-Leg Sitting Position

The open-leg sitting position provides a medium base of support and is most commonly used when firing from a forward slope.

Three-Point Sling. The open-leg sitting position is a higher profile position than the other sitting positions because the torso is more erect, making it more advantageous to assume when wearing a combat load. To assume the open-leg sitting position with the three-point sling—

- Stand erect and face the target at approximately a 45-degree angle to the target.
- Grasp the pistol grip with your firing hand.
- Grasp the rail system/fore grip/grip pod with the support hand under the sling in a position that provides maximum bone support and stability of the weapon.
- Bend at the knees and, while elevating the muzzle, drop to one knee and roll back onto your buttocks.
- Extend your legs to an open position.
- Pull back on both your firing and support hands to place the Service rifle butt into your firing shoulder so that the sights are level with your eyes.
- Place your cheek firmly against the stock to obtain a firm stock weld.
- Place your support elbow on your support-side leg below the inside of the knee. Lower your firing elbow to the inside of your firing-side knee. Slight inboard muscular tension of the legs may be required to support this position.

See figure 6-21.

Figure 6-21. Open-Leg Sitting Position with Three Point Sling.

To adjust for elevation—

- Move your support hand rearward or forward on the rail system. Moving your hand rearward elevates the muzzle.
- Move the fore grip/grip pod up in your grasp to raise the muzzle and move it down in your grasp to lower the muzzle.

To adjust for a minor cant in the Service rifle—

- Rotate the rail system left or right by rotating the pistol grip left or right.
- Rotate the fore grip/grip pod left or right.

Loop Sling. To assume the open-leg sitting position with the loop sling—

- Stand erect and face the target at a 30 to 40 degree angle to the right of the line of fire.
- Place your feet approximately shoulder-width apart.
- Place your support hand over the sling from the left side and under the rail system.
- Bend at the knees while breaking your fall with your firing hand.
- Push backward with your feet to extend your legs and place your buttocks on the ground.
- Place your support elbow on your support-side leg below the knee.
- Grasp the Service rifle butt with your firing hand and place it into your shoulder pocket.
- When the Service rifle butt becomes seated, grasp the pistol grip firmly with your firing hand.
- Lower your firing elbow to the inside of your firing-side knee.
- Place your cheek firmly against the stock to obtain a firm stock weld.
- Move your support hand to a location under the rail system that will provide maximum bone support and stability of the weapon. Slight muscle tension of the legs may be required to support the Service rifle.

See figure 6-22.

Figure 6-22. Open-Leg Sitting Position with Loop Sling.

Support in the Sitting Position

As time and the combat situation permit, the Marine should seek cover and use support to assist in stabilizing the Service rifle sight(s). A sitting position—

- Can be used to fire over the top of cover when mobility is not as critical.
- Can be comfortably assumed for a longer period of time than a kneeling position and can conform to higher cover when a prone position cannot be used.

Support the sitting position by placing the rail system/fore grip/grip pod, the forearm, or the magazine on or against support (see fig. 6-23). If the Service rifle is resting on support, the Marine may not need to stabilize the weapon by placing his support or firing elbows on his legs (see fig. 6-24).

Figure 6-23. Sitting Position: Forearm Against Support.

Figure 6-24. Sitting Position: Service Rifle on Support.

Kneeling Position

The kneeling position is—

- Quick to assume.
- Easy to maneuver from.
- Normally assumed after the initial engagement has been initiated from a standing position.
- Can easily be adapted to available cover.

A tripod is formed by the support-side foot, firing-side foot, and firing-side knee when the Marine assumes the kneeling position, providing a stable foundation for shooting. The kneeling position also presents a higher profile to facilitate a better field of view as compared to the prone and sitting positions.

The kneeling position can be assumed by either moving forward or dropping back into position, depending on the combat situation. For example, it may be necessary to drop back into position to avoid crowding cover or to avoid covering uncleared terrain. To move forward into the kneeling position, the Marine steps forward toward the target with his support foot and kneels down on his firing knee. To drop back into the kneeling position, the Marine leaves his support foot in place, steps backward with his firing foot and kneels down on his firing knee.

High-Kneeling Position

The high-kneeling position can be assumed with both the three-point and loop slings.

Three-Point Sling. To assume the high-kneeling position with the three-point sling—

- Stand erect and face the target at approximately a 45-degree angle to the target.
- Grasp the pistol grip with your firing hand.
- With your support hand, grasp the rail system/fore grip/grip pod under the sling in a position that provides maximum bone support and stability of the weapon.
- Bend at the knees and while elevating the muzzle, drop until the firing-side knee is on the deck.
- Pull back on both your firing and support hands to place the Service rifle butt into the firing shoulder so that the sights are level with your eyes.
- Place your cheek firmly against the stock to obtain a firm stock weld.
- Keep your firing-side ankle straight, with the toe of your boot in contact with the ground and curled under by the weight of your body. Place the firing-side portion of your buttocks on your firing-side heel, making solid contact.
- Place your support-side foot forward to a point that allows your shin to be vertically straight. Your support-side foot should be flat on the ground.
- Lean slightly forward and place the flat part of your upper support arm, just above the elbow, on your support-side knee so that it is in firm contact with the flat surface formed on top of your bent knee. The point of the support elbow extends just past the support-side knee.
- Bend your firing elbow to provide the least muscular tension possible, and lower it to a natural position.

See figure 6-25 on page 6-24.

To adjust for elevation—

- Move your support hand rearward to elevate the muzzle.
- Move the fore grip/grip pod up in the grasp to raise the muzzle and down in the grasp to lower the muzzle.

To adjust for a minor cant in the Service rifle—

- Rotate the rail system left or right by rotating the pistol grip left or right.
- Rotate the fore grip/grip pod left or right.

Figure 6-25. High-Kneeling Position with Three-Point Sling.

Loop Sling. To assume the high-kneeling position with the loop sling—

- Stand with your feet approximately shoulder-width apart and face the target approximately 45 degrees to the right of the line of fire.
- Step forward with your support-side foot toward the target.
- Place your support hand under the rail system.
- Kneel down on your firing-side knee so your firing-side lower leg is approximately parallel to the gun-target line.
- Keep your firing-side ankle straight, with the toe of your boot in contact with the ground and curled under by the weight of your body.
- Place the firing-side portion of your buttocks on your firing-side heel, making solid contact.
- Place your support-side foot forward until your shin is vertically straight. Your support-side foot should be flat on the ground, since it will be supporting the majority of your weight.
- Place the flat part of your upper support arm, just above the elbow, on your support-side knee so that it is in firm contact with the flat surface formed on top of your bent knee. This means that the point of your support elbow will extend just slightly past the support-side knee.
- Lean slightly forward into the sling for support.
- Grasp the Service rifle butt with your firing hand and place the butt of the Service rifle into the pocket of your shoulder.
- Grasp the pistol grip with your firing hand.
- Bend your firing elbow to provide the least muscular tension possible and lower it to a natural position.
- Relax your weight forward and place your cheek firmly against the stock to obtain a correct stock weld.
- Move your support hand to a location under the rail system, providing maximum bone support and stability for the weapon.

See figure 6-26.

Variations of the High-Kneeling Position. A variation of the high-kneeling position addresses elevation and is used to fire over cover and provide elevation when stability and a low profile are

Figure 6-26. High-Kneeling Position with Loop Sling.

also needed. The Marine assumes the kneeling position by dropping down to one knee and keeping the upper body upright. The buttocks are not in contact with the foot and the firing-side upper leg remains vertical. The support-side elbow does not rest on the knee, but may rest on cover. This is the same lower body position assumed when told to take a knee.

See figure 6-27.

Figure 6-27. High-Kneelling Modified for Elevation.

Another variation of the high-kneeling position is the two-knee position (see fig. 6-28 on page 6-26). To assume the two-knee variation, drop both knees onto the deck. The toes should be curled to get into and out of the position quickly. Depending on the need for observation of

Figure 6-28. Two-Knee Kneeling Position.

the enemy, the buttocks may or may not rest on the heels. Resting the buttocks on the heels will provide additional stability to the position, but less mobility.

Medium-Kneeling Position

The medium-kneeling position (see fig. 6-29) is also referred to as the bootless kneeling position. Assume the medium-kneeling position in the same way as the high-kneeling position, with the exception of the firing-side foot. The firing-side ankle is straight and the foot is stretched out with the bootlaces in contact with the ground. The buttocks are in contact with the heel of the firing-side foot.

Figure 6-29. Medium-Kneeling Position.

Low-Kneeling Position

The low-kneeling position (see fig. 6-30) is most commonly used when firing from a forward slope. Assume the low-kneeling position in the same way as the high-kneeling position, with the exception of the placement of the firing-side foot. Turn the firing-side ankle so the outside of the foot is in contact with the ground and the buttocks are in contact with the inside of the foot.

Figure 6-30. Low-Kneeling Position.

Support in the Kneeling Position

As time and the combat situation permit, the Marine should seek cover and use support to assist stabilizing the Service rifle sight(s). When the prone position cannot be used because of the height of the support, the kneeling position may be appropriate. The kneeling position—

- Provides additional mobility over the prone position.
- Allows shooting from all sides and from cover of varying sizes.

This position can be altered to maximize the use of cover or support by assuming a variation of the kneeling position (i.e., high, medium, or low).

In the kneeling position, the Marine must not telegraph (i.e., broadcast) his position behind the cover with his knee. When shooting around the sides of cover, the Marine should strive to keep his firing knee in line with his support foot so that he does not reveal his position to the enemy.

Support the position by placing the rail system/fore grip/grip pod, the forearm, or the magazine on or against support. In addition, the position (e.g., a knee, the side of the body) can rest against support (see fig. 6-31 on page 6-28).

If the Service rifle is resting on support, the Marine may not need to stabilize the weapon by placing his support elbow on his knee (see fig. 6-32 on page 6-28).

Standing Position

The standing position is the quickest position to assume and the easiest from which to maneuver. It allows for greater mobility than other positions and—

- Is often used for immediate combat engagement.
- Is supported by the Marine's legs and feet, providing a small area of contact with the ground.

> ***Note:*** The body's center of gravity is high above the ground; therefore, maintaining balance is critical in this position.

Figure 6-31. Kneeling Position Using Support.

Figure 6-32. Kneeling Position with Service Rifle on Support.

The standing position will be the default position for most initial and close-range engagements. Its primary benefits are mobility and observation. A properly built standing position will enable the Marine to rapidly and effectively engage multiple close-range targets, while permitting 360-degree movement. This position may be assumed with either the three-point or parade sling.

Three-Point Sling

To assume the standing position for quick, close engagement with the three-point sling—

- Square your body to the target.
- Spread your feet apart to a comfortable distance with your support foot slightly in front of your firing foot. This distance may be wider than shoulder width.

- Distribute your weight evenly over both feet and hips and shift your balance forward slightly to reduce recovery time and increase the stability of your hold. Your legs should be bent slightly for balance.
- Grasp the pistol grip with your firing hand.
- With your support hand, grasp the rail system/fore grip/grip pod under the sling in a position that provides maximum bone support and stability of the weapon. If grasping the rail system, your support hand will be under them, with your thumb on the outboard side of them. The magazine must be on the inside of your support arm.
- Bring the Service rifle sights up to eye level instead of lowering your head to the sights and place your cheek firmly against the stock. Ensure that your head is erect so that your aiming eye can look through the rear sight aperture.
- Pull back on both your firing and support hands to place the Service rifle butt into your firing shoulder so that the sights are level with your eyes.
- Hold your firing elbow in a natural position.

See figure 6-33.

Figure 6-33. Standing Position with Three-Point Sling.

To adjust for elevation—

- Move your support hand rearward or forward on the rail system. Moving the hand rearward elevates the muzzle.
- Move the fore grip/grip pod up in the grasp to raise the muzzle and move it down in the grasp to lower the muzzle.

To adjust for a minor cant in the Service rifle—

- Rotate the rail system left or right by rotating the pistol grip left or right.
- Rotate the fore grip/grip pod left or right.

To gain additional stability for the position as time and distance allow, move your support arm rearward and brace the back of your upper arm against your equipment or torso. This will increase bone support to assist with holding the Service rifle up, rather than relying strictly on muscular tension.

Parade Sling

During training, the standing position is fired using a parade sling so that the body provides the balance and support of the weapon to stabilize the sights. To assume the standing position with a parade sling—

- Stand erect.
- Face approximately 90 degrees to the right of the line of fire.
- Place your feet approximately shoulder width apart.
- Distribute your weight evenly over your feet and hips. Your legs should be straight, but your knees should not be locked.
- Place your support hand under the rail system in a position to best support and steady the Service rifle. Your support-side triceps can rest against your torso, but cannot rest or be supported by equipment mounted on the cartridge belt.
- Grasp the pistol grip with your firing hand.
- Place the Service rifle butt into your shoulder pocket.
- Position your support elbow across your upper torso. The majority of the Service rifle's weight is held with your support arm resting naturally against your upper torso and should be supported by bone structure, not muscle.
- Hold your firing elbow in a natural position.
- Bring the Service rifle sights up to eye level instead of lowering your head to the sights. Ensure that your head is erect to allow you to look straight through the sights. Eye relief will normally be increased in the standing position because of the placement of the Service rifle butt and your head being held more erect.
- Place the stock firmly against your cheek in the same place each time to ensure consistency from shot to shot.

See figure 6-34.

Support in the Standing Position

As time and the combat situation permit, the Marine should seek cover and use support to assist stabilizing the Service rifle sight(s). The standing position can be adapted to cover, while still providing greater mobility and observation of the enemy than other positions. The standing position can effectively be used either behind high cover (e.g., window, over a wall) or narrow cover (e.g., tree, telephone pole).

To use artificial support from the standing position, the Marine will lean his body forward or against support to stabilize the weapon and the position. He will support the position by placing the rail system/fore grip/grip pod, his forearm, or the magazine on or against support. The position (e.g., the side of the body) can also rest against support (see fig. 6-35).

Figure 6-34. Standing Position with Parade Sling.

Figure 6-35. Standing Position Using Support.

Considerations for Using Support

Supports are foundations for positions and positions are foundations for the Service rifle. To maximize the support that the position provides, the firing position should be adjusted to fit or conform to the shape of the cover. Elements of a sound firing position, such as balance and stability, must be incorporated and adjusted to fit the situation and type of cover. A supported firing position should—

- Minimize exposure to the enemy.
- Maximize the stability of the Service rifle.
- Provide protection from enemy observation and fires.

The Marine can use any available support (e.g., logs, rocks, sandbags, walls) to stabilize his firing position. The surrounding combat environment dictates the type of support and position used.

Adjusting the Firing Position

The type of cover can dictate which firing position (i.e., prone, sitting, kneeling, or standing) will be the most effective. For example, the Marine's height in relation to the height of the cover aids in the selection of a firing position. The firing position selected should be adjusted to fit the type of cover to—

- *Provide stability.* The firing position should be adjusted to stabilize the Service rifle sights and allow the management of recoil to recover on target.
- *Permit mobility.* The firing position should be adjusted to permit lateral engagement of dispersed targets and movement to other cover.
- *Allow observation.* The firing position should be located to allow observation of the area/enemy while minimizing exposure to the enemy.

> ***Note:*** The firing position should be adjusted to fit the type of cover by adjusting the support hand, pocket of shoulder, firing elbow, stock weld, and/or grip of the firing hand in support of the Service rifle or position.

- *Keep the entire body behind cover.* The Marine should minimize exposure of any part of his body to fire and be aware of any body part that may extend beyond the cover (e.g., the head, firing elbow, knees).
- *Shoot from the right or left side of cover.* To minimize exposure and maximize the cover's protection, if possible, a right-handed Marine should shoot from the right side of cover, and a left-handed Marine should shoot from the left side (see fig. 6-36). However, if a right-handed Marine must fire from the left side of cover, he shoots right-handed, but adjusts his position behind cover (see fig. 6-37).

Firing Over the Top of Cover

Firing over the top of cover provides a wider field of view and lateral movement. When firing over the top of cover, the position can be supported and stabilized by resting the rail system or the support forearm on the cover (see fig. 6-38). The Marine should keep as low a profile as possible and the Service rifle should be as close to the top of cover as possible.

Figure 6-36. Firing from Right Side of Cover.

Figure 6-37. Firing from Left Side of Cover.

Figure 6-38. Firing Over the Top of Cover.

Maintaining Muzzle Awareness

When firing over the top of cover, the Marine must remember that the sights are higher than the barrel and remain aware of the location of his muzzle. The RCO/iron sight optic line of sight is approximately 3 inches above the line of fire from the muzzle of the weapon. This differential must be considered because while a clear line of sight through the RCO/sight to target is acquired, the muzzle of the weapon may not be above any obstructions; such as the cover, intervening objects (e.g., turret shield, vehicle part), or microterrain directly in front of the Service rifle. Therefore, the Marine must maintain a position that ensures the muzzle is high enough to clear the cover (e.g., window sill, top of wall) as he obtains sight picture on the target (see fig. 6-39).

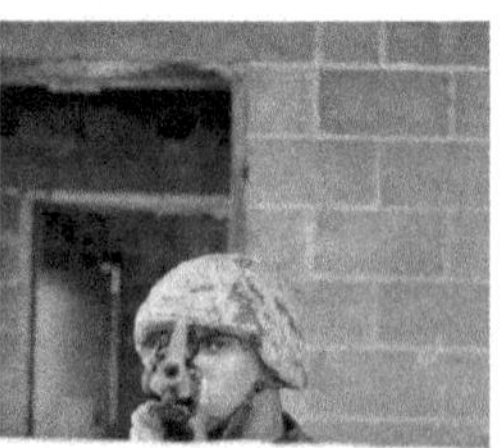

Muzzle not clear

Muzzle clear

Figure 6-39. Clearing Cover with the Muzzle.

Clearing the Ejection Port

Ensure the cover does not obstruct the ejection port. If the ejection port is blocked, the obstruction can interfere with the ejection of the spent cartridge case and cause a stoppage.

Resting the Magazine on Support

The bottom, front, or side of the Service rifle magazine can rest on or against support to provide additional stability (see fig. 6-40 and fig. 6-41).

CAUTION

The back of the magazine should not be pulled back against support because it can cause a stoppage by not allowing a round to feed from the magazine.

Using the Support Hand for Stability on Cover

The support hand should be used to help stabilize both the firing position and the Service rifle to enable the Marine to maintain sight alignment and sight picture.

Figure 6-40. Magazine on Support.

Figure 6-41. Front of Magazine Against Support.

The forearm or support hand can contact the support to stabilize the weapon (see fig. 6-42).

Figure 6-42. Forearm Resting on Cover.

The Service rifle rail system can rest on the support, but the barrel cannot (see fig. 6-43). Placement of the support hand on the rail system may need to be adjusted forward or backward

Figure 6-43. Rail System Resting on Cover.

to accommodate the cover and the additional support provided by the Service rifle resting on the cover. If the rail system is resting on the cover, the support hand can pull down on the rail system to further stabilize the weapon.

The body weight can be shifted forward to stabilize the position against cover (see fig. 6-44).

Figure 6-44. Shifting Body Weight Into Cover.

Firing from Specific Types of Cover

Effective cover allows the Marine to engage enemy targets while protecting himself from enemy fire. Several types of cover provide support, protection, and concealment and do not interfere with target engagement. The Marine must adapt firing positions to the type of cover that is available.

Window

The Marine can establish a supported or unsupported position from a window.

The Marine can establish a position back from the opening of the window so that the muzzle does not protrude and interior shadows provide concealment without providing a silhouette to the enemy (see fig. 6-45 on page 6-38).

When additional stability is needed, the Marine can establish a position by placing the Service rifle rail system or his forearm/hand in the V formed by the side and bottom of the windowsill. A drawback to this technique is that the muzzle of the weapon and the Marine may be exposed to view (see fig. 6-46, on page 6-38, and fig. 6-47 on page 6-39).

Figure 6-45. Window Position.

Figure 6-46. Window Position: Rail System on Support.

Vehicle

In many combat situations, particularly in urban environments, a vehicle may be the best form of cover. When using a vehicle for cover, the engine block provides the most protection from small-arms fire. The Marine should establish a position behind the front wheel so that the engine

Figure 6-47. Window Position: Forearm/Hand on Support.

block is between him and the target (see fig. 6-48 and fig. 6-49 on page 6-40). From this position, the Marine can fire over, under, or around the vehicle. This is a very effective position that can be used for larger vehicles that are higher off the ground.The Marine can also establish additional support for the Service rifle by positioning himself behind the doorjamb (i.e., frame of door) and placing the Service rifle against the V formed by the open door and doorframe (see fig. 6-50 on page 6-40). From this position, the Marine can fire over the hood of the vehicle while using the engine block for protection.

Figure 6-48. Firing Around Front of Vehicle.

Figure 6-49. Firing Over Front of Vehicle.

Figure 6-50. Establishing a Supported Position in a Vehicle.

At the back of the vehicle, only the axle and the wheel provide cover. If the Marine must shoot from the back of the vehicle, he must position himself directly behind the wheel as much as possible (see fig. 6-51).

Figure 6-51. Using the Back of a Vehicle for Cover.

THIS PAGE INTENTIONALLY LEFT BLANK

CHAPTER 7

Effects of Weather

All weather conditions have a physical and psychological effect on Marines. Through proper training, Marines can develop the confidence required to reduce the physical and psychological effects that are caused by weather.

Wind, temperature, precipitation, and light can affect the trajectory of the bullet, so Marines must use techniques to offset these effects.

Physical and Psychological Effects of Weather

Wind

Marines can shoot effectively in windy conditions if they apply a few basic techniques and develop the proper mental attitude. The Marine can combat the wind in the following ways:

- Make subtle changes to the basic firing positions, such as increasing muscular tension, to reduce movement of the Service rifle sights.
- Select a more stable firing position.
- Seek support to stabilize the Service rifle.
- Hold the shot and apply the fundamentals during a lull in the wind.

Temperature

Extreme Heat

In extreme heat, the Marine may experience rapid fatigue. Heat can cause muscle cramps, heat exhaustion, heat stroke, blurred vision, and reduced concentration levels that result in inaccurate shooting. During extreme heat—

- Good physical condition and increased fluid intake can help to offset any adverse effects.
- Sweat running into the eyes can cause irritation and make it difficult to see the sights.
- Ground mirages can cause a target to appear indistinct and to drift from side to side. Heat waves or mirages may also distort the target shape.

To overcome the effects of heat and accurately engage a target, the Marine should maintain a center mass hold. Heat waves and mirages appear as waves and can be observed through an optic to determine wind speed and direction. The bigger the wave, the less strong the wind; the smaller the wave, the greater the speed of the wind. In desert or urban conditions where there are no trees, the Marine should observe the dust or mirage to gauge wind speed and direction.

Extreme Cold

Extreme cold can affect the Marine's ability to concentrate. If the Marine's hands are numb, he will have difficulty holding a frigid Service rifle and executing effective trigger control. To protect the hands in a cold environment, the Marine should wear arctic mittens or gloves. To operate the Service rifle while wearing arctic mittens or gloves, the Marine depresses the trigger guard plunger to open the trigger guard. This allows easier access to the trigger (see fig. 7-1).

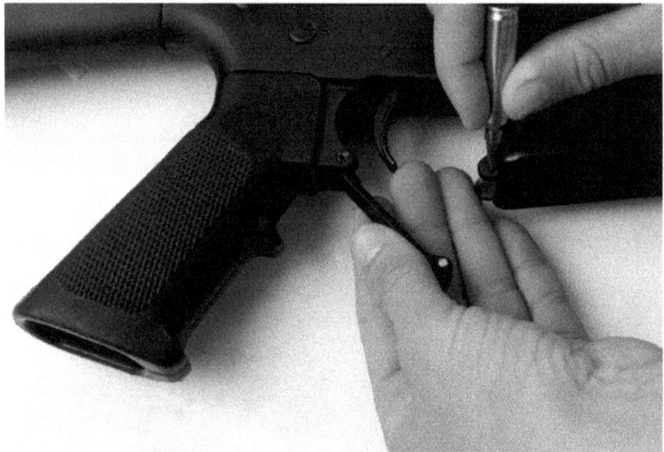

Figure 7-1. Open Trigger Guard.

Precipitation

Precipitation (e.g., rain, snow, hail, sleet) can affect target engagement, the Marine's comfort level, and the Marine's ability to concentrate. The amount and type of precipitation can obscure or completely hide the target, reducing the Marine's ability to establish an accurate sight picture. In addition, precipitation collecting on a rear sight aperture can make it difficult to establish sight alignment and sight picture. Sights should be protected as much as possible during periods of precipitation. Since it is easy to lose concentration when wet and uncomfortable, the proper attire should be worn to reduce the effects of precipitation on the Marine.

Light

Light conditions—

- Can affect each Marine differently.
- Can change the appearance of a target.
- Can affect range estimation, visual acuity, or the placement of the tip of the front sight on the target.

By maintaining a center mass hold, the effects of light can be reduced.

Bright Light

Bright light conditions exist under a clear blue sky with no fog or haze present to filter the sunlight. Bright light can make a target appear smaller and farther away. As a result, it is easy to overestimate range. Maintaining a center mass hold, regardless of how indistinct the target appears, ensures the best chance for an effective shot.

Overcast

An overcast condition exists when a solid layer of clouds block the sun. The amount of available light changes as the overcast thickens. Overcast conditions can make a target appear larger and closer. As a result, it is easy to underestimate range. During a light overcast, both the target and the Service rifle sights appear very distinct, making it easy to establish sight alignment. As the overcast thickens, it becomes difficult to distinguish the target from its surroundings.

Haze

Hazy conditions exist when fog, dust, humidity, or smoke are present. Hazy conditions can make a target appear indistinct making it difficult to establish sight picture.

Physical Effects of Wind on the Bullet

The weather condition that presents the greatest problem to shooting is the wind because it affects a bullet's trajectory. The effect of wind on the bullet as it travels down range is referred to as deflection. The wind deflects the bullet laterally in its flight to the target (see fig. 7-2).

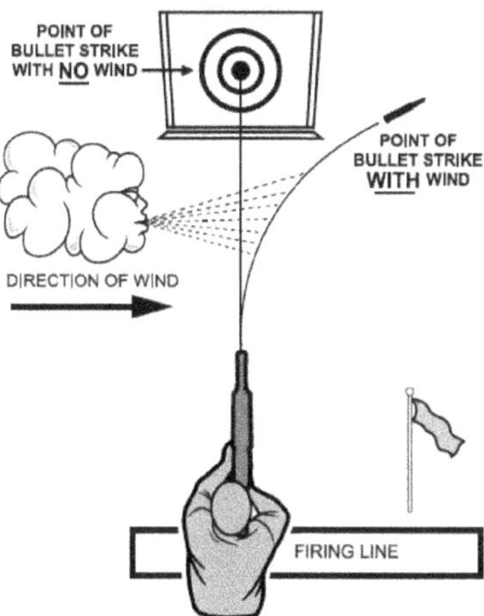

Figure 7-2. Deflection of a Bullet.

The bullet's exposure time to the wind determines the amount the bullet is deflected from its original trajectory. Deflection increases as the distance to the target increases. There are three factors that affect the amount of deflection of the bullet:

- *Velocity of the wind.* The greater the velocity of the wind, the more the bullet will be deflected.
- *Range to the target.* As the distance to the target increases, the speed of the bullet slows, allowing the wind to have a greater effect on shot placement.
- *Velocity of the bullet.* A bullet with a high muzzle velocity will not be affected by the wind as much as a bullet with a low muzzle velocity.

Windage Adjustments for the Rifle Combat Optics

Offset Aiming

Unlike the adjustable iron sights on the Service rifle, the RCO should not be adjusted for a wind change. The windage turrets on the RCO should only be adjusted during zeroing. For wind corrections during firing, offset aiming is employed so that a hold into the direction of the wind will produce the desired result. Offset aiming for windage is covered in chapter 9.

Reading a Mirage for Wind Direction

When observing an area through the RCO, a mirage can be read to determine wind direction. Mirage refers to the heat waves or the reflection of light through the layers of air of different temperature and density as seen by the human eye on a warm, bright day. A mirage—

- Can be seen best on bright sunny days over unbroken terrain.
- Can be seen with an RCO (i.e., optic) on all but the coldest days, with varying degrees of clarity depending on light and temperature.
- Can generally be seen off of surfaces that generate heat in warm conditions, such as streets and roofs in an urban environment.

A mirage is particularly valuable when reading no value winds, where the mirage gives the appearance of moving straight up with no lateral movement. This is called either a boiling mirage or just a boil. If the mirage is boiling, the effective wind velocity is zero.

The general appearance of the mirage waves can aid in determining wind direction. A lateral moving mirage is indicated by the heat wave becoming horizontal instead of vertical, as in a boiling mirage. For example, the waves will move from left to right if the wind is blowing left to right. The stronger the wind, the more flattened the mirage becomes. When aimed in on the target, the mirage is indicating the wind at the target, which has the least effect on the bullet's trajectory. If time permits, focus on a mirage halfway or three-quarters the distance to your target to get a true sense of the wind direction, because the wind at this distance will have the greatest effect on the bullet.

Windage Adjustments to Compensate for Wind Effects for Iron Sights

Iron sights are moved to compensate for the wind's effect on the strike of a bullet. The velocity and direction of the wind in relationship to the bullet must be determined to offset the wind's effects. If Marines can classify wind values and determine velocity within 5 miles per hour (mph), they can effectively engage targets in windy conditions.

Wind Direction

Determine wind direction by observing the direction the vegetation is moving and by feeling the wind blow against the body.

Wind Value Classifications

Winds are classified according to the direction that they are blowing from in relation to the direction of fire. The clock system indicates wind direction and value (see fig. 7-3). Winds can be classified as half value, full value, or no value. The target is always located at 12 o'clock.

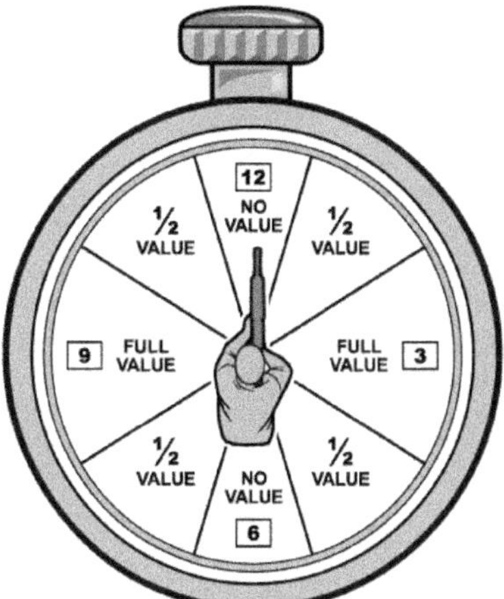

Figure 7-3. The Clock System.

Wind Velocity

There are three methods to determine wind velocity.

Flag Method. The flag method is used as a training tool on the KD range to learn the observation method for estimating wind velocity. To estimate wind velocity in mph—

- Estimate the angle created between the flagpole and the flag in degrees.
- Divide the angle by 4 to estimate wind velocity in mph (see fig. 7-4 on page 7-6).

> *Note:* Information provided by the flag method should be based on a typical flag used in the pits when dry, measuring 5 feet wide, and tapering down to 3 feet wide and 18 feet long.

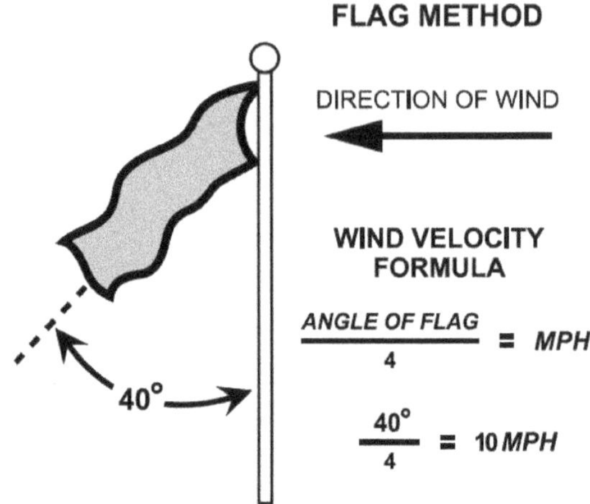

Figure 7-4. Flag Method.

After identifying wind direction, wind classification, and wind velocity, windage adjustments that are required for the bullet to strike the target must be estimated. To estimate windage adjustments, match the wind velocity, wind direction, and range to the target to the information in the charts in table 7-1 and table 7-2, on page 7-8, to determine the correct number of clicks to apply to the windage knob. Once the number of windage clicks is determined, turn the windage knob, causing the rear sight aperture to move into the direction of the wind.

Note: In training, target dimensions are as follows:

A target: 4 ft x 6 ft target face; aiming black = 12 inches; 4-ring = 24 inches; 3-ring = 36 inches.

D target: 6 ft x 6 ft target face; aiming black = 26 inches wide x 19 inches tall; 4-ring = 34 inches wide x 37 inches high; 3-ring = 60 inches wide x 51 inches tall.

B-MOD target: 6 ft x 6 ft target face; aiming black = 40 inches tall x 20 inches wide; 4-ring = 40 inches; 3-ring = 60 inches.

E target: 40 inches tall x 20 inches wide.

Table 7-1. Windage Click Chart for Flag Method for M4.

WINDAGE CLICK CHART FLAG METHOD										
RANGE FLAG ANGLES Wind speed is determined by the angle of the flag. The different speeds at each angle can be approximated based on how fast the flag flutters at each angle.	20° 5 MPH		40° 10 MPH		60° 15 MPH		80° 20 MPH		90° 25 MPH	
RANGE YARDS	WIND VALUE		WIND VALUE		WIND VALUE		WIND VALUE		WIND VALUE	
	FULL	HALF	FULL	HALF	FULL	HALF	FULL	HALF	FULL	HALF
200	1	1	2	1	3	1	4	2	5	2
300	2	1	4	2	7	3	9	4	11	5
500	4	2	8	4	12	6	16	8	20	10

Table 7-2. Windage Click Chart for Flag Method for M16A4.

WINDAGE CLICK CHART FLAG METHOD										
RANGE FLAG ANGLES Wind speed is determined by the angle of the flag. The different speeds at each angle can be approximated based on how fast the flag flutters at each angle.	20° 5 MPH		40° 10 MPH		60° 15 MPH		80° 20 MPH		90° 25 MPH	
RANGE YARDS	WIND VALUE		WIND VALUE		WIND VALUE		WIND VALUE		WIND VALUE	
	FULL	HALF	FULL	HALF	FULL	HALF	FULL	HALF	FULL	HALF
200	2	1	3	1	5	2	6	3	8	4
300	3	1	6	3	10	5	13	6	16	8
500	6	3	12	6	18	9	24	12	30	15

Pointing Method. In an operational environment where no flag is present, the pointing method may be used. The pointing method determines the value of the wind at your location and involves—

- Facing directly into the wind and executing a right face, drop a light object from shoulder height.
- Pointing at the object and determining the angle to the object from the side of your body.
- Taking the angle that your arm is out from your body and dividing it by 4 to determine the wind velocity in mph.

The accuracy of this method depends on the object that is being dropped. The calculated wind velocity will be more for dry grass than for wet grass, since wet grass is heavier.

Observation Method. The observation method is the primary method that is used to estimate wind velocity and direction in a tactical situation. This method teaches Marines to relate the effect that a given wind condition has on the natural surroundings. The following are guidelines used during the observation method:

- 5 mph winds can be detected by the presence of a slight wind or by drifting smoke; can be felt lightly on the face; and keep tree leaves in a constant motion.
- 10 mph winds raise dust and loose paper.
- 15 mph winds cause small trees to sway.
- 20 mph winds cause large trees to sway.

Once wind direction, wind classification, and wind velocity have been identified, refer to the windage charts in tables 7-1 and 7-2 for the correct windage adjustments that are needed to enable the bullet to strike the target. Match the wind velocity, wind direction, and range to the target to the information in these figures to estimate the correct number of clicks to apply to the windage knob.

Windage Adjustment Calculations

This method uses a simple mathematical formula to determine how many clicks of windage should be applied to the Service rifle to compensate for the effects of wind. This formula is only good for the Service rifle, and is adjusted in accordance with full wind value.

The M16A4 rifle has a windage scale that is gauged in 1/2 inch per 100 meters/yards of range per click on the rear sight windage knob.

The M4 carbine's windage scale is gauged in 3/4 inch per 100 meters/yards of range per click on the rear sight windage knob.

$$\frac{\text{range X velocity in mph}}{\text{range constant}} = \text{clicks for full value wind}$$

Formula for the M16A4 Rifle. The range constant is based on the range to the target but it depends on the type of ammunition and weapon system fired because these factors affect the velocity of the round. The range constant for M855 ammunition fired from an M16A4 rifle is as follows:

- If the range to the target is 200 to 400 meters/yards, the range constant is 5.
- If the range to the target is 500 to 700 meters/yards, the range constant is 4.

The range that the formula uses is measured in 100-meter/yard increments. At a range of 300 meters/yards to the target, a **3** would be entered for the range in the formula.

When adjusting the windage on the Service rifle, the rear sight aperture must always be moved into the direction that the wind is blowing. For example, if the wind is blowing from the right, the rear sight aperture must be moved right. See figure 7-5 for an example of the M16A4 rifle formula.

A 10 mph wind is blowing from 9 o'clock.

The range to the target is 500 meters/yards; therefore, if—

Range (R)= 5

Velocity (V) = 10

500 meters/yards

Range constant = 4

Then—

$\frac{R \times V}{4} = \frac{5 \times 10}{4} = \frac{50}{4}$ = 12.5 or 13 clicks left on the windage knob of an M16A4 rifle

Figure 7-5. Example of M16A4 Rifle Formula.

Formula for the M4 Carbine. The range constant for M855 ammunition fired from an M4 carbine is as follows:

- If the range to the target is 200 meters/yards, the range constant is 10.
- If the range to the target is 300 to 400 meters/yards, the range constant is 7.
- If the range to the target is 500 to 700 meters/yards, the range constant is 6.

See figure 7-6 for an example of the M4 carbine formula.

Physical Effects of Temperature and Precipitation on the Bullet and Service Rifle

Temperature

Extreme changes in temperature promote fluctuation in the Service rifle's chamber pressure that are caused by changes in the propellant's temperature. In cold weather, as Service rifle chamber pressure decreases, the bullet exits the muzzle at a lower velocity, impacting the target below the point of aim. In extreme heat, the Service rifle's chamber pressure increases, causing the bullet to exit the muzzle at a higher velocity and impact the target above the point of aim.

A 10 mph wind is blowing from 9 o'clock.

The range to the target is 500 meters/yards; therefore, if—

Range (R) = 5

Velocity (V) = 10

500 meters/yards

Range constant = 6

Then—

$\frac{R \times V}{6} = \frac{5 \times 10}{6} = \frac{50}{6}$ = 8.3 or 8 clicks left on the windage knob of an M4 carbine

Figure 7-6. Example of M4 Carbine Formula.

> ***Note:*** Hot air is less dense than cool air and provides less resistance to the bullet, allowing the bullet to travel faster and experience less deflection from the wind. Cold air is dense and provides the bullet with more resistance, causing the bullet to travel slower and experience greater deflection from the wind. Ammunition should be protected from direct exposure to the sun to avoid changes in propellant temperature.

Once the Service rifle is zeroed, a change in temperature of 20 degrees or more can cause the bullet to strike above or below the point of aim. Therefore, if the temperature changes 20 degrees or more, the Marine should rezero the Service rifle.

If the Service rifle is exposed to below freezing temperatures, it should not be brought into a warm location immediately. Condensation may form on and in the Service rifle and freeze if reexposed to the cold. Ice can form on the—

- Inside of the Service rifle, causing it to malfunction.
- Rear sight aperture due to condensation, making it impossible to acquire sight picture.

Precipitation

Freezing rain and other types of precipitation can—

- Make the Service rifle difficult to handle.
- Foul the Service rifle and cause stoppages.
- Build up in the barrel or compensator and cause erratic shots.

Care should be taken to keep the barrel and muzzle free of water. Therefore, the Service rifle should be carried weak-side sling arms (i.e., muzzle down) to keep the moisture out of the bore.

If the Service rifle has been submerged in water, ensure that the bore is drained before firing. To drain the bore, pull the charging handle slightly to the rear and hold for a few seconds while the muzzle points down. Once the barrel has been drained, turn the Service rifle muzzle up to allow the water to drain out of the stock.

THIS PAGE INTENTIONALLY LEFT BLANK

CHAPTER 8

Zeroing the Service Rifle

To be combat effective, it is essential for the Marine to know how to zero his Service rifle. Zeroing is adjusting the sights on the weapon to cause the shots to impact where the Marine aims. This must be accomplished while compensating for the effects of weather and the range to the target. It is critical that Marines can zero their Service rifles and make the sight adjustments required to engage targets accurately.

Elements of Zeroing

There are five basic elements involved in zeroing a Service rifle: line of sight, aiming point, centerline of the bore, trajectory, and range (see fig. 8-1).

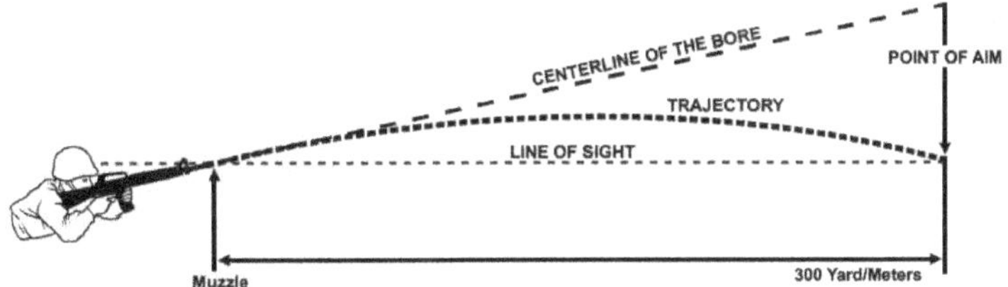

Figure 8-1. Elements of Zeroing.

Line of Sight

The line of sight is a straight line, which begins with the Marine's eye and proceeds through the center of the sight(s) to a point of aim on a target.

Aiming Point

The aiming point is the precise point where the tip of the front sight post or RCO reticle pattern is placed in relation to the target.

Centerline of the Bore

Centerline of the bore is an imaginary straight line beginning at the chamber end of the barrel, proceeding out of the muzzle, and continuing indefinitely.

Trajectory

In flight, a bullet does not follow a straight line, but travels in a curve or arc, called trajectory. Trajectory is the path a bullet travels to the target. As the bullet exits the muzzle, it immediately starts to fall because of gravity. To compensate for that fall, the sights are set to elevate the muzzle. As the bullet exits the muzzle it intersects the line of sight, because the sights are above the muzzle. As the bullet travels farther, it reenters the line of sight to intersect with it again (see fig. 8-2).

> *Note:* When mounted on the Service rifle, the optic center of the RCO sits approximately 2.75 inches above the centerline of the bore.

Range

Range is the distance from the Service rifle muzzle to the target. Range estimation is discussed in chapter 11.

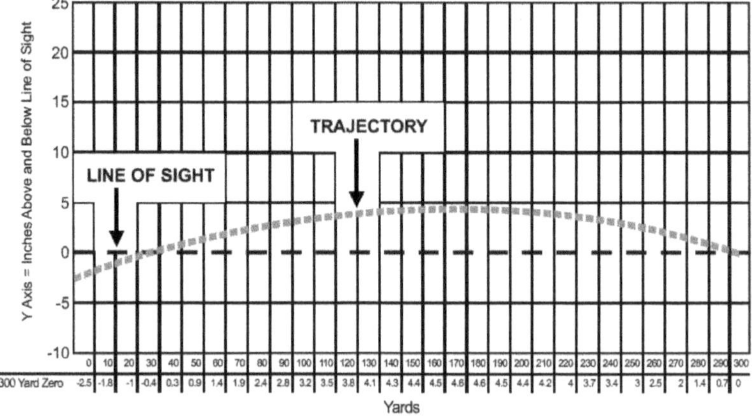

Figure 8-2. Trajectory 0 to 300 Meters.

Types of Zeroes

Zero

Zero for the Rifle Combat Optic

A zero for the RCO is the elevation and windage settings required to place a single shot or the center of a shot group in a predesignated location on a target at 100 meters/yards, from a specific firing position, under ideal weather conditions. Holds for wind and elevation are discussed in chapter 9.

Zero for Backup Iron Sights

The elevation and windage settings required to place a single shot or the center of a shot group, in a predesignated location on a target at a specific range, from a specific firing position, under specific weather conditions.

True Zero

A true zero is the elevation and windage settings (BUIS) that are required, to place a single shot or the center of a shot group, in a predesignated location on a target at a specific range, from a specific firing position, under ideal weather conditions.

Battlesight Zero Backup Iron Sights

Battlesight zero is the—

- Elevation and windage settings that are required (BUIS) to place a single shot or the center of a shot group in a predesignated location on a target at 300 meters/yards under ideal weather conditions.
- Sight settings placed on the Service rifle for combat. In combat, the Service rifle's BZO setting will enable engagement of point targets from 0 to 300 meters/yards in a no-wind condition.

Windage and Elevation Adjustments Using the Rifle Combat Optic

Mechanical Adjustments

The RCO is optically centered when it leaves the manufacturer. Windage and elevation adjusters are used to zero the optic. The adjusters can be moved with a coin, bladed screwdriver, or the extractor rim of the 5.56-mm casing. Adjustment increments are 1/3 inch per click at 100 meters/yards.

> ***Note:*** This means that nine clicks are required to move the strike of the bullet 1 inch at 33 meters/36 yards; three clicks are required to move the strike of the bullet 1 inch at 100 meters/yards; and one click is required to move the strike of the bullet 1 inch at 300 meters/yards.

To mechanically adjust the RCO—

- Move the windage adjuster clockwise to move the strike of the round to the right or move it counterclockwise to move the strike of the round to the left (see fig. 8-3).
- Move the elevation adjuster clockwise to move the strike of the round up or move it counterclockwise to move the strike of the round down (see fig. 8-4).

Adjusting the optic is a precision exercise and should be performed by—

- Moving the adjusters slowly.
- Counting the number of clicks based on the distance you need to move the strike of the round.

> *Note:* The Marine should be able to hear and feel the audible tactile clicks. Once the RCO is zeroed, do not remove the adjuster caps.

Figure 8-3. Service Rifle Combat Optic Windage Adjuster.

Figure 8-4. Service Rifle Combat Optic Elevation Adjuster.

CAUTION

Forcing the adjusters to their extremes will render the optic useless. If you encounter resistance and continue adjusting, the resistance will cease and the adjuster will turn freely again, damaging the RCO.

If the adjusters are turned counterclockwise beyond the surface of the adjuster housing, the adjuster could be removed. Do not attempt to remove the adjuster.

Bullet Drop Compensator

The RCO is calibrated to accommodate bullet drop. The entire reticle pattern is a bullet drop compensator with designated aiming points to compensate for trajectory of the 5.56-mm round at ranges of 100 to 800 meters/yards. This feature eliminates the need for mechanical elevation adjustments on the Service rifle. To compensate for range to the target, the following aiming points are used when zeroing the RCO (see fig. 8-5):

- Hold the tip of the chevron center mass on a target at 100 meters/yards.
- Hold the tip of the red post center mass on a target at 300 meters/yards. This same hold is used at 33 meters/36 yards, because the BZO target is scaled to represent a target at 300 meters/yards.

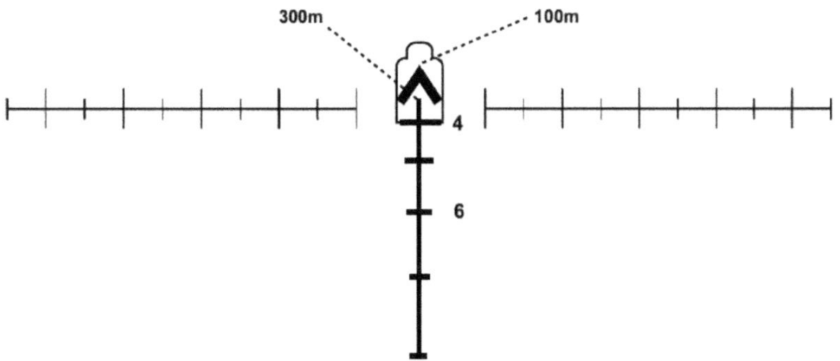

Figure 8-5. RCO Aiming Points.

Zeroing Process for the Rifle Combat Optic

Zeroing for the RCO is conducted at 100 meters/yards. To zero the Service rifle, you will need to laser boresight and prezero the sight setting for the RCO.

Laser Boresight

Laser boresight the RCO according to the instructions in Technical Manual 10471A-12P/1, *Operator's Manual for the Laser Boresight System (LBS)*. If a laser boresight system is not available, a prezero sight setting can be established on the RCO at 33 meters/36 yards. A prezero sight setting is performed on a short range to get the Marine's shots on paper in preparation for zeroing at 100 meters/yards.

Prezero Sight Setting for the Rifle Combat Optic

A zero established at 33 meters/36 yards is not as accurate as a zero established at 100 meters/yards. Therefore, a prezero sight setting is established at 33 meters/36 yards and the weapon is zeroed at 100 meters/yards. To establish a prezero sight setting on the weapon—

- Place a universal zeroing target at a range of 33 meters/36 yards. (A sample of this target is found in appendix A). To be accurate, the target must be placed exactly 33 meters/36 yards from the muzzle of the Service rifle. When firing the RCO, the chevron of the reticle is placed in the triangle-shaped portion of the target, with an equal amount of white space on either side of the chevron. This places the tip of the red post at the 0 line of the target grid. A 1-inch designated impact area for shot groups is shaded in grey (see fig. 8-6).
- Place a piece of tape over the fiber optic to create a finer aiming point on the reticle. This provides a finer aiming point on the top of the reticle.
- Establish a stable, supported prone firing position (e.g., sandbag, assault pack, bipod).
- Use the 300-m aim point on the reticle as follows:
  –The tip of the red post should be placed at the 0 line of the target grid.
  –The red chevron should be placed in the triangle-shaped portion of the target with an equal amount of white space on either side of the chevron (see fig. 8-7).
- Fire three rounds to obtain a shot group.
- Triangulate the shot group to find the center.
- Determine the vertical and horizontal distance in inches from the center of the shot group to the center of the target. Each grid of the target equals 1/2 inch. Perform the following:
  –Adjust the reticle to move the center of the shot group to the desired point of impact.
  –Adjust nine clicks to move the strike of the round 1 inch at 33 meters/36 yards for both windage and elevation.

> ***Note:*** Sight adjustments on the RCO may require additional shots to allow the reticle to seat properly.

- Fire three rounds to obtain a shot group.
- Adjust the reticle to move the center of the shot group to the desired point of impact.
- Fire four rounds to confirm the prezero sight setting.
- Adjust the reticle as necessary to move the center of the shot group to the desired point of impact.

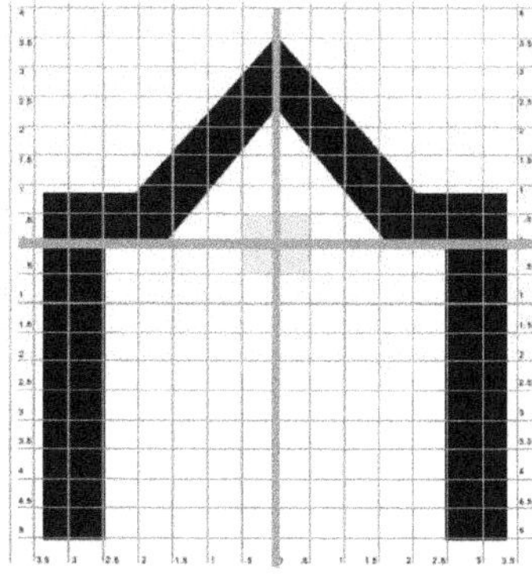

Figure 8-6.
Universal Battlesight Zero Target.

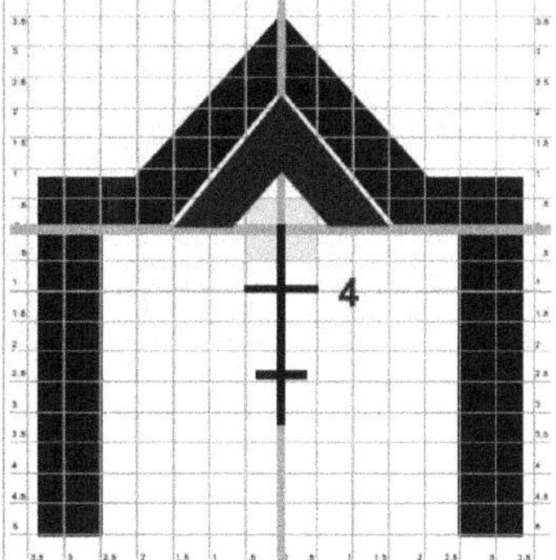

Figure 8-7.
Rifle Combat Optic Sight Picture.

Rifle Combat Optic Zeroing

Zeroing the RCO is conducted at 100 meters/yards.

> *Note:* A zero is not established by simply obtaining a prezero sight setting. A zero established at 33 meters/36 yards is not as accurate as a zero established at 100 meters/yards.

To zero the RCO—

- Place a suitable target with an aiming point that is 4 inches in diameter and in contrast with the background (e.g., V ring of an A target) at a range of 100 meters/yards and determine an aiming point (see fig. 8-5). Use the 100-m aim point on the reticle: tip of the chevron center mass on the target.
- Establish a stable, supported prone firing position (e.g., sandbag, assault pack, bipod).
- Fire three rounds to obtain a shot group.
- Triangulate the shot group to find the center.
- Determine the vertical and horizontal distance in inches from the center of the shot group to the center of the target.
- Adjust the reticle to move the center of the shot group to the desired point of impact. Three clicks moves the strike of the round 1 inch at 100 meters/yards for both windage and elevation.
- Fire three rounds to obtain a shot group.
- Adjust the reticle to move the center of the shot group to the desired point of impact.
- Fire four rounds to confirm the zero. The Service rifle is considered zeroed when a shot group is inside the 4-inch aiming area of the target.

Backup Iron Sight

The BUIS system consists of a rear sight aperture, a rear sight elevation drum, and a rear sight windage knob.

Backup Iron Rear Sight

The rear sight consists of a single sight aperture that is used for all firing situations (see fig. 8-8).

Figure 8-8. Backup Iron Rear Sight.

Backup Iron Sight Rear Elevation Drum

The rear sight elevation drum is used to adjust the sight for a specific range to the target. The elevation drum is indexed as shown in figure 8-9. Each number on the drum represents a distance from the target in 200-meter/yard increments, out to 600 meters/yards. To adjust for range to the target, rotate the elevation drum so that the desired setting is aligned with the index on the left side of the sighting system.

> ***Note:*** A hasty sight setting is the setting placed on the rear sight elevation drum to engage targets beyond 300 meters. Hasty sight settings for ranges of 400 to 600 meters/yards are applied by rotating the rear sight elevation drum to the number that corresponds with the engagement distance of the enemy.

Backup Iron Windage Knob

The windage knob, located on the right side of the sighting system, is used to adjust the strike of the round right or left. The windage knob is marked with an arrow and the letter **R** that shows the direction the strike of the round is being moved (see fig. 8-10).

Figure 8-9. Backup Iron Elevation Drum.

Figure 8-10. Backup Iron Windage Knob.

To move the strike of the round to the right, rotate the windage knob clockwise in the direction of the arrow marked **R**.

To move the strike of the round to the left, rotate the windage knob counterclockwise.

Backup Iron Sight Zeroing Process

A prezero sight setting is established on the BUIS at 33 meters/36 yards using the same procedures as those for the RCO (see pg. 8-6). Ideally, zeroing of the BUIS should be conducted at the same time that the RCO is zeroed. The recommended sequence of events is as follows:

- Mount the BUIS on the last groove of the rail system and flip it up. Set the elevation drum on the BUIS at **3**.
- Establish a prezero sight setting on the BUIS at 33 meters/36 yards as follows:
  - –The front sight post is moved to make elevation adjustments.
  - –The BUIS windage knob is moved to make lateral adjustments. Three clicks of the windage adjustment will move the strike of the round on the target approximately 1/2 inch at 33 meters/36 yards.
- Leave the elevation drum on **3**. Flip the BUIS down. Mount the RCO on the rail system just forward of the BUIS. Establish a prezero sight setting on the RCO at 33 meters/36 yards.
- Zero the RCO at 100 meters/yards (see pg. 8-6).

Iron Sights Sighting System

The sighting system of the Service rifle consists of a front sight post, two rear sight apertures, an elevation knob, and a windage knob. Scales of the sighting system can be applied accurately to both yard and meter measurements. For example, a rear sight elevation setting of **6/3** can be used for 300 and 600 meters/yards.

Front Sight

The front sight post is used to adjust for elevation. The front sight consists of a square rotating sight post with a four-position spring-loaded detent (see fig. 8-11).

Figure 8-11. Front Sight.

To adjust for elevation, use a pointed instrument or the tip of a cartridge to depress the detent and rotate the front sight post (see fig. 8-12). To raise the strike of the bullet, rotate the post clockwise (i.e., in the direction of the arrow marked **UP**) or to the right. To lower the strike of the bullet, rotate the post counterclockwise (i.e., in the opposite direction of the arrow) or to the left.

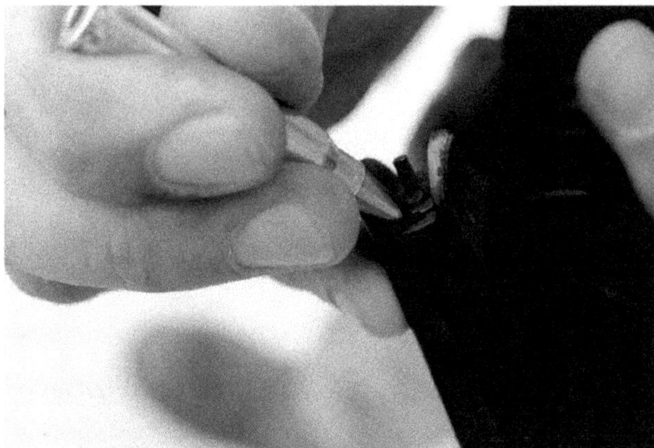

Figure 8-12. Front Sight Adjustment.

Rear Sight

The rear sight consists of two sight apertures: an elevation knob and a windage knob (see fig. 8-13). The large aperture marked **0-2** is used for target engagement during limited visibility when a greater field of view is desired or for engagements of targets closer than 200 meters/ yards. The unmarked, or small aperture, is used for zeroing and normal firing situations.

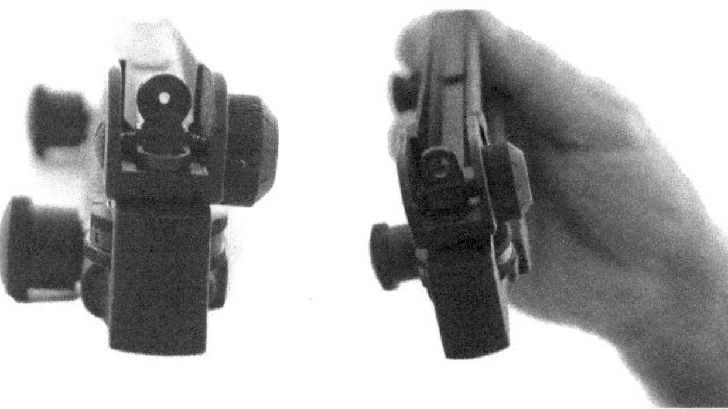

Figure 8-13. Rear Sight.

Elevation Knob

The rear sight elevation knob is used to adjust the sight for a specific range to the target. The elevation knob is indexed as shown in figure 8-14. Each number on the knob represents a distance from the target in 100-meter/yard increments, out to 600 meters/yards. To adjust for range to the target, rotate the elevation knob so that the desired setting is aligned with the index on the left side of the receiver.

Figure 8-14. Elevation Knob.

If the elevation knob is turned so that—

- The number **6/3** aligns with the elevation index line, the **3** indicates 300 meters/yards (see fig. 8-15).
- The numbers **4** and **5** are aligned with the elevation index line. It places the elevation at 400 and 500 meters/yards, respectively.

Figure 8-15. Rear Sight Elevation Knob Set for 300 Meters/Yards.

If a clockwise rotation is continued, the number **6/3** appears for the second time on the elevation index line and indicates a 600-meter/yard elevation. When the rear sight elevation knob is set on **6/3** for 600 meters/yards, there will be a considerable gap (i.e., about 1/4 inch) between the rear sight housing and the upper receiver (see fig. 8-16).

When time permits, the Marine may place a hasty sight setting on his Service rifle to engage targets beyond the BZO range of the Service rifle. Hasty sight settings for ranges of 400 to

Figure 8-16. Rear Sight Elevation Knob Set for 600 Meters/Yards.

600 meters/yards are applied by rotating the rear sight elevation knob to the number that corresponds with the engagement distance of the enemy.

To achieve a hasty sight setting, the Marine dials the appropriate range numeral on the rear sight elevation knob that corresponds to the range to the target. For example, if the rear sight elevation knob is set at **6/3** and a target appears at 500 meters/yards, rotate the knob to the **5** setting.

Upon completion of firing with a hasty sight setting for extended ranges, return the rear sight to the BZO setting.

Windage Knob

The windage knob is used to adjust the strike of the round to either the right or left. The windage knob is marked with an arrow and the letter **R** that show the direction the strike of the round is being moved (see fig. 8-17).

To move the strike of the round to the right, rotate the windage knob clockwise in the direction of the arrow marked **R**.

To move the strike of the round to the left, rotate the windage knob counterclockwise.

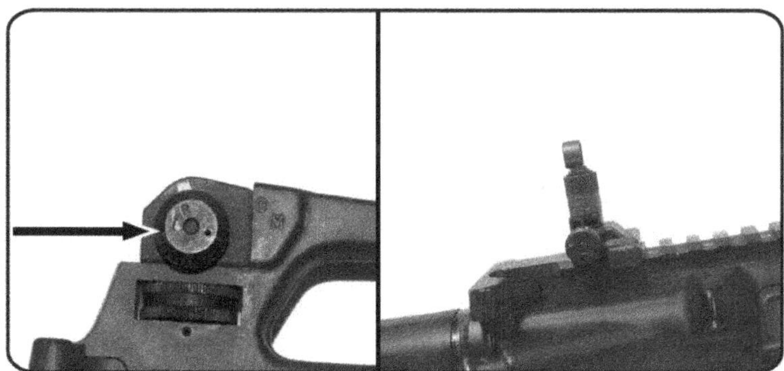

Figure 8-17. Windage Knob.

Windage and Elevation Rules for Iron Sights

Moving the front sight post, elevation knob, or windage knob one graduation or notch is referred to as "moving one click on the sight." The windage and elevation rules define how far the strike of the round will move on the target for each click of front and rear sight elevation or rear sight windage for each 100 meters/yards of range to the target.

Front Sight Elevation Rule

One click of front sight elevation adjustment will move the strike of the round on target approximately—

- 1 3/8 inches for every 100 meters/yards of range from the target for the M16A4.
- 1 7/8 inches for every 100 meters/yards of range from the target for the M4 carbine.

Rear Sight Elevation Rule

One click of the rear sight elevation adjustment will move the strike of the round on the target approximately—

- 1/2 inch for every 100 meters/yards of range from the target for the M16A4.
- 3/4 inch for every 100 meters/yards of range from the target for the M4 carbine.

Windage Rule

One click of windage adjustment will move the strike of the round on the target approximately—

- 1/2 inch for every 100 meters/yards of range from the target for the M16A4.
- 3/4 inch for every 100 meters/yards of range from the target for the M4 carbine.

Iron Sights Zeroing Process

During the zeroing process, all elevation adjustments are made on the front sight post. Once a BZO is established, the front sight post should never be moved, except when rezeroing the Service rifle. Zeroing is conducted at a range of 300 meters/yards. To prepare a Service rifle for zeroing, the Service rifle sights must be adjusted to the initial sight settings.

> ***Note:*** The rear sight elevation knob is used for dialing in the range to the target.

Initial Sight Settings

Initial sight settings are those settings that serve as the starting point for initial zeroing and all sight adjustments. If there is already a BZO established on the Service rifle, the zeroing process may begin by using the previously established BZO sight settings. The following subparagraphs identify how to set the sights to initial sight settings.

Front Sight Post

To set the front sight post to the initial sight setting, depress the front sight detent and rotate the front sight post until the base of the front sight post is flush with the front sight housing.

Rear Sight Elevation Knob

To set the elevation knob at the initial sight setting, perform the following:

- Rotate the rear sight elevation knob counterclockwise until the moveable rear sight housing is bottomed out on the upper receiver.

  Note: Once bottomed out, the rear sight elevation knob should be six clicks counterclockwise from **6/3**.

- Rotate the rear sight elevation knob clockwise until the number **6/3** aligns with the index mark located on the left side of the upper receiver (see fig. 8-18).

Figure 8-18. Elevation Knob Set at 6/3.

Windage Knob

To set the windage knob to the initial sight setting, rotate the windage knob until the index line located on the top of the large rear sight aperture aligns with the centering on the windage index scale located on the moveable base of the rear sight assembly (see fig. 8-19).

Figure 8-19. Aligning Index Line.

Iron Sights for Prezero Sight Setting

Zeroing is conducted at a range of 300 meters/yards. If a 300-meter/yard range is not available, a prezero sight setting can be established at a reduced range of 33 meters/36 yards. This does not constitute a BZO. The process at this short range allows the Marine to "get a group" on paper in preparation for firing at 300 meters/yards. When a Service rifle is zeroed for 300 meters/yards, the bullet crosses the line of sight twice. First, it crosses the line of sight at 33 meters/36 yards and then again farther down range at 300 meters/yards as it reenters the line of sight (see figs. 8-20 and 8-21).

> ***Note:*** The mathematics in figures 8-20 and 8-21 are calculated based on yards.

Therefore, a Service rifle's prezero sight setting is established at a distance of 33 meters/36 yards and the same BZO will be effective at 300 meters/yards.

A Service rifle that only has a prezero sight setting established, not a BZO, should not be used in battle. A BZO established at 300 meters/yards is considered to be much more accurate than the 33-meter/36-yard prezero sight setting. This is because of the minor inconsistencies that

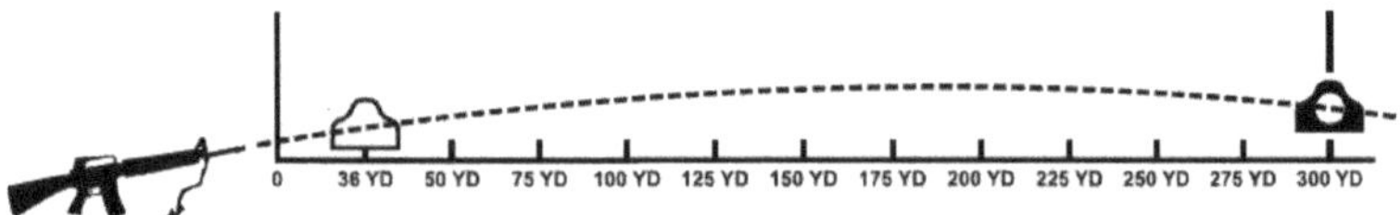

Figure 8-20. Bullet Crossing the Line of Sight Twice.

the Marine can apply at 33 meters/36 yards that will be greatly multiplied at 300 meters/yards.

To establish a prezero sight setting at 33 meters/36 yards—

- Place a universal zeroing target (see app. A) at a range of 33 meters/36 yards (see fig. 8-22). To be accurate, the target must be placed exactly 33 meters/36 yards from the muzzle of the Service rifle. When firing iron sights, the tip of the front sight post is placed in the square-shaped portion of the target with an equal amount of white space on either side of the post; the tip of the post is placed on the **0** line of the grid target (see fig. 8-23). A 1-inch designated impact area for shot groups is shaded in grey.
- Establish a stable supported prone firing position (e.g., sandbag, assault pack, bipod).
- Fire three rounds to obtain a shot group.
- Triangulate the shot group to find the center as follows:
  - –Determine the vertical distance in inches from the center of the shot group to the center of the target. Each grid of the target equals 1/2 inch. Make elevation adjustments on the front sight post to move the center of the shot group to the center of the target.
  - –Determine the horizontal distance from the center of the shot group to the center of the target. Each grid of the target equals 1/2 inch. Make lateral adjustments on the windage knob to move the center of the shot group to the center of the target.
- Fire three rounds to obtain a shot group.
- Make adjustments to move the center of the shot group to the desired point of impact.

- Fire four rounds to refine the zero.
- Adjust the sights as necessary to move the center of the shot group to the desired point of impact.

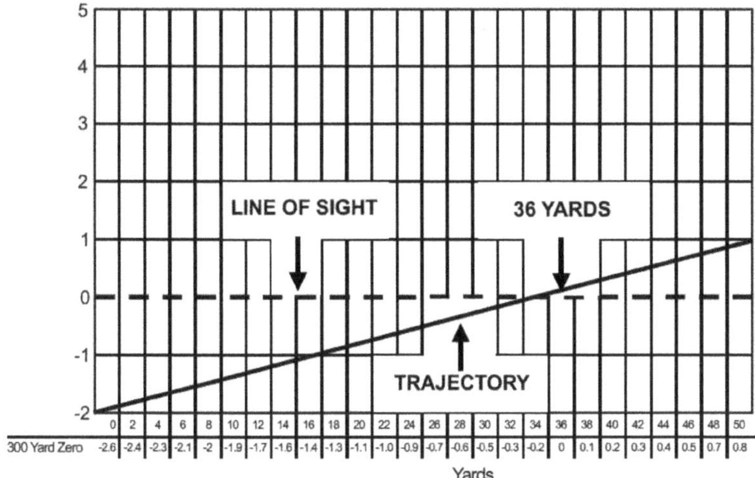

Figure 8-21. Trajectory 0 to 50 Yards.

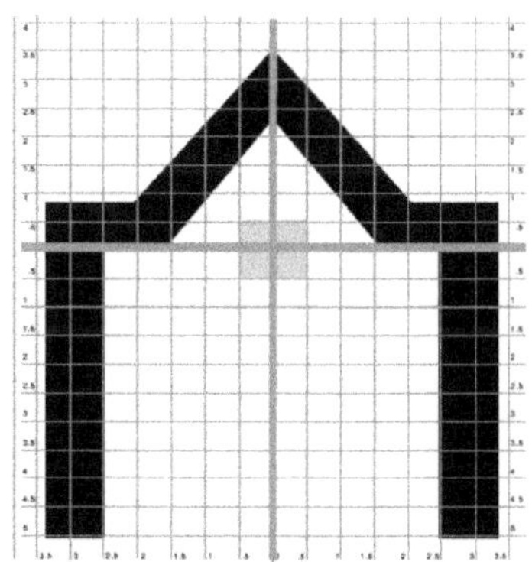

Figure 8-22. Universal Zeroing Target.

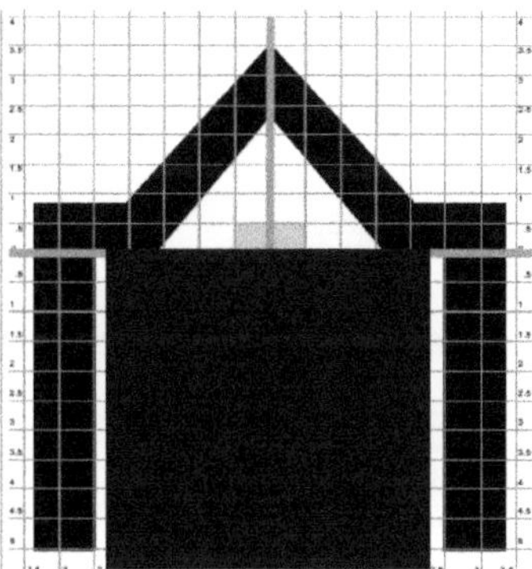

Figure 8-23. Iron Sights Sight Picture.

The Zeroing Process

Zeroing is performed at 300 meters/yards on the D target. To zero the Service rifle at a range of 300 meters/yards, perform the following steps:

- Fire a three-shot group.
- Triangulate the shot group to find the center as follows:
  - –Determine the vertical distance in inches from the center of the shot group to the center of the target. Make elevation adjustments on the front sight post to move the center of the shot group to the center of the target.
  - –Determine the horizontal distance from the center of the shot group to the center of the target. Make lateral adjustments on the windage knob to move the center of the shot group to the center of the target.
- Fire three rounds to obtain a shot group.
- Make adjustments to move the center of the shot group to the desired point of impact.
- Fire four rounds to confirm the sight setting.
- Determine the value and direction of the wind and remove the number of clicks added to the windage knob, if necessary, to compensate for current wind conditions. This becomes the BZO setting for the Service rifle.

Factors Causing a Battlesight Zero Reconfirmation

Marines are responsible for maintaining a BZO on their Service rifles at all times. Factors that can influence the BZO of a Service rifle to change on a daily basis include atmospheric conditions, humidity, and temperature. If operating in a combat environment, Marines should confirm their BZO as often as possible. To confirm a BZO, the Marine can begin the zeroing process by using the previously established BZO sight settings rather than placing the sights at initial sight setting. Maintenance, temperature, climate, ammunition, ground elevation, and uniform will cause a BZO to need to be reconfirmed.

Maintenance

It is possible for the BZO to change if ordnance personnel perform maintenance on a Service rifle. If maintenance was performed, it is critical that the Service rifle be rezeroed as soon as possible.

Temperature

An extreme change in temperature (i.e., 20 degrees or more) will cause the elevation BZO to change. Changes in temperature cause chamber pressure to increase when hot and decrease when cold, causing shots to impact the target high in hot temperatures and low in cold temperatures.

Climate

Changing climates (e.g., moving from a dry climate to a tropical climate) can cause fluctuation in air density, moisture content, temperature, or barometric pressure and affect the Service rifle's BZO.

Ammunition

Inconsistencies in the production of ammunition lots can change a Service rifle's BZO.

Ground Elevation

Drastic changes in ground elevation can create changes in air density, moisture content, temperature, or barometric pressure, affect the Service rifle's BZO.

Uniform

If Marines zero their Service rifles in utility uniform and fire in full battle gear, their BZOs can change. The wearing of full battle gear changes eye relief, placement of the Service rifle in the shoulder pocket, and the way the Service rifle is supported on the rail system. Marines must establish their BZOs while wearing the uniform and equipment that they will be wearing while engaging targets.

Factors Affecting Accuracy of Battlesight Zero

Anything that the Marine changes from shot to shot affects the accuracy of his BZO, which influences the accuracy of shot placement. The following factors, when applied inconsistently, diminish the accuracy of a BZO:

- Any of the seven factors—support hand, Service rifle butt in the pocket of the shoulder, grip of the firing hand, firing-side elbow, stock weld, muscular control, or breathing.
- Hold stability.
- Sling tension.
- Trigger control.
- Sight picture.

THIS PAGE INTENTIONALLY LEFT BLANK

CHAPTER 9

Offset Aiming for Windage and Elevation

To engage a target during combat, the Marine may be required to aim his Service rifle at a point on the target other than center mass. This is known as offset aiming. Offset aiming involves adjusting sight picture to compensate for the distance and size of the target to account for wind conditions and range to the target or elevation.

Offset Aiming for Wind

The strike of the round can be affected by wind. For wind corrections during firing, offset aiming is used. Unlike the adjustable iron sights on the Service rifle, the RCO should not be adjusted for a wind change. The windage adjusters on the RCO should only be adjusted during zeroing. A hold into the direction of the wind, based on wind speed, will enable accurate engagement of a target. Offset aiming must be used to compensate for the strike of the round when wind is a factor.

A hold for windage should be based on something that can be visually seen and estimated with some uniformity, such as the width of a body. For example, the width of a body is considered approximately 19 inches wide; half a body width is approximately 9.5 inches. Holds will vary based on the wind speed, range to the target, and the weapon system. The RCO reticle pattern centered on the edge of the target into the wind is a hold of approximately 9.5 inches, which is considered a hold of half a body width (see fig. 9-1).

Direction of Wind

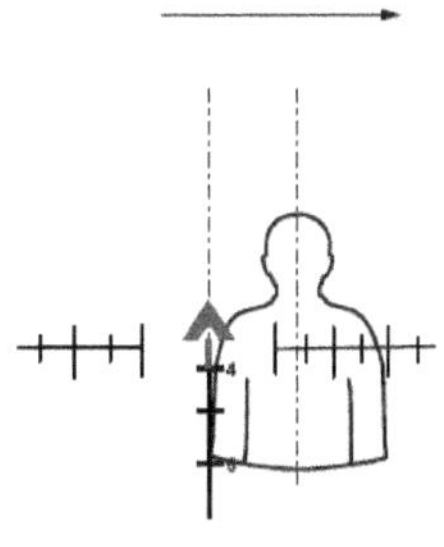

Hold of Half Body Width

Figure 9-1. Rifle Combat Optic Holds for Wind.

The conditions of Service rifle fire in combat may not permit mechanical adjustments of iron sights. Offset aiming must be used to compensate for the strike of the round when wind is a factor and, in the case of iron sights, when there is no time to adjust the sights (see fig. 9-2).

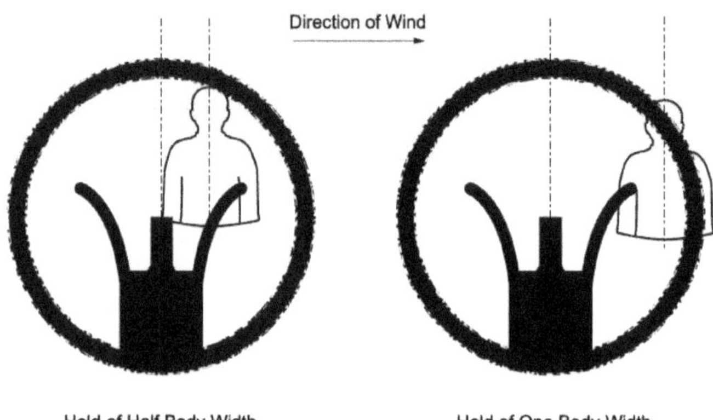

Figure 9-2. Iron Sights Holds for Wind.

Guidelines for Applying Offset Aiming

The following general guidelines apply when using offset aiming to compensate for a full value wind:

- For distances of 200 meters/yards, with light to medium winds (e.g., 5–15 mph), the aiming point is center mass. A hold is not required.
- For distances of 300 meters/yards, with light winds (e.g., 5 mph), the aiming point is center mass. A hold is not required.
- For distances of 200 meters/yards, with heavy winds (e.g., 20 mph), hold half a body width (i.e., on the edge of the target) into the wind.
- For distances of 300 meters/yards, with 10 mph winds, hold half a body width (i.e., on the edge of the target) into the wind.
- For distances of 200 meters/yards, with strong winds (e.g., 25 mph), hold one body width into the wind.
- For distances 300 meters/yards and medium to heavy winds (e.g., 15–20 mph), hold one body width into the wind.
- For distances 300 meters/yards and strong winds (e.g., 25 mph), hold one and a half body widths into the wind.
- For distances of 400 meters/yards, with light winds (e.g., 5 mph), hold half a body width on the edge of the target into the wind.
- For distances of 400 meters/yards, with 10 mph winds, hold one body width into the wind.
- For distances of 500 meters/yards, with light winds (e.g., 5 mph), hold one body width into the wind.

Tables 9-1 and 9-2 present the actual windage adjustments in inches when firing the M4 and M16A4, respectively.

Table 9-1. Windage Chart in Inches for the M4 Carbine.

Wind Speed	5 mph		10 mph		15 mph		20 mph		25 mph	
	Wind Value									
Range	Full	Half	Full	Half	Full	Half	Full	Half	Full	Half
200	3	1	5	2	8	4	10	5	19	9
300	6	3	13	6	18	9	25	12	32	16
400	12	6	22	11	32	16	42	21	54	27
500	20	10	40	20	60	30	81	40	101	50

Table 9-2. Windage Chart in Inches for the M16A4 Carbine.

Wind Speed	5 mph		10 mph		15 mph		20 mph		25 mph	
	Wind Value									
Range	Full	Half	Full	Half	Full	Half	Full	Half	Full	Half
200	2	1	5	2	7	3	9	4	11	5
300	5	2	11	5	16	8	22	11	27	13
400	8	4	18	8	26	13	36	18	44	22
500	17	8	35	17	52	26	69	34	87	43

Offset Aiming for Wind During Training

During training, the A, D, and B-MOD targets are fired upon. The dimensions of the targets must be known in order to equate windage holds in body widths (i.e., 19 inches =1 body width). Figure 9-3 presents the targets with a body superimposed over each one in order to gauge body width.

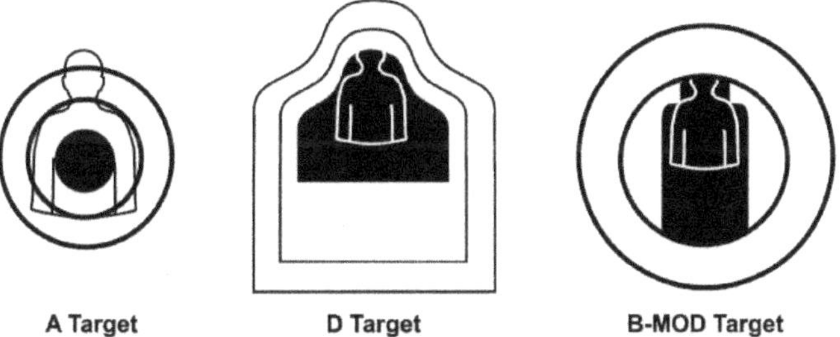

Figure 9-3. Training Targets Equated to Body Width.

Offset Aiming to Compensate for Range to Target or Elevation

External Ballistics

Ballistics is the science of projectiles and their effects. External ballistics—

- Are important to understand as they relate to trajectory.
- Are what a projectile is doing while in flight.
- Apply to the flight of the projectile after it exits the bore. As soon as the projectile exits the bore, velocity begins to decrease because there is no longer any force to speed up or maintain the speed of the round. Air drag immediately begins to effect the round, slowing it down. As the projectile ends its contact with the lands and grooves of the barrel, the gas-tight seal that was behind the projectile is broken. The expanding gas behind the bullet exits along with the bullet and is expelled around the bullet and bore. The bullet will not be stable when it exits the bore due to lack of constriction provided by the barrel and by the exiting gasses. The bullet will wobble in flight (i.e., precession) until gyroscopic action stabilizes it. Gyroscopic stabilization keeps the round pointed in the same direction as the barrel and causes the round to resist changes in orientation.

Trajectory 0 to 300 Meters/Yards

The following applies once the round exits the barrel when trajectory of the round is at 0 to 300 meters/yards.

- Upon exiting the barrel, the round must travel up to meet the line of sight because of the height of the sights over the bore (see fig. 9-4). The round travels forward and up until it reaches its maximum height of trajectory or maximum ordinate, which will be at roughly two-thirds of its target distance. At that point, the round begins its descent onto the target. If the Service rifle is properly zeroed for 300 meters/yards, the trajectory (or path of the bullet) will rise approximately 4 1/2 inches (i.e., 11 centimeters) above the line of sight at a distance of approximately 175 yards or 160 meters. At other distances, the strike of the bullet will be less than 4 1/2 inches above the point of aim.

> ***Note:*** Only at 33 meters/36 yards and 300 meters/yards does the point of impact coincide with the point of aim.

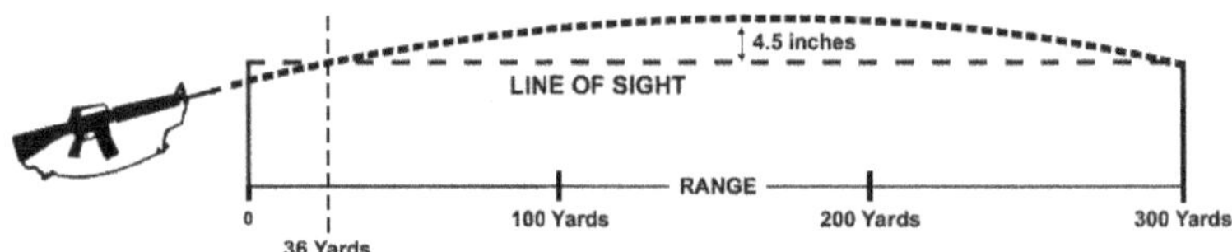

Figure 9-4. Trajectory for 300-Meter/Yard Battlesight Zero.

- The bullet will rise roughly 5 inches above the line of sight between 165 to 180 meters/yards for an M16A4 rifle firing M855 ammunition that was zeroed at 300 meters/yards. At that point, the round begins its descent onto the target (see fig. 9-5 and fig. 9-6 on page 9-6).
- A Service rifle zeroed at 300 meters/yards will cross the line of sight on its upward path at 33 meters/36 yards and will cross the line of sight again on its downward path at 300 meters/yards. Because point of aim and point of impact are the same, that is why zeroing is conducted at 300 meters/yards (see fig. 9-5).
- At 100 meters/yards, the 300-meter/yard zero will be 3 inches high, so the Marine must aim lower (see fig. 9-5).
- If only a portion of the target is visible (e.g., the head of an enemy), the trajectory of the bullet may need to be taken into consideration when firing at a distance less than 300 meters/yards. If the Marine does not consider trajectory, he can shoot over the top of the target if the target is small and at a distance less than 300 meters/yards (see fig. 9-5).

> *Note:* The ballistics data in figure 9-5 is calculated based on yards.

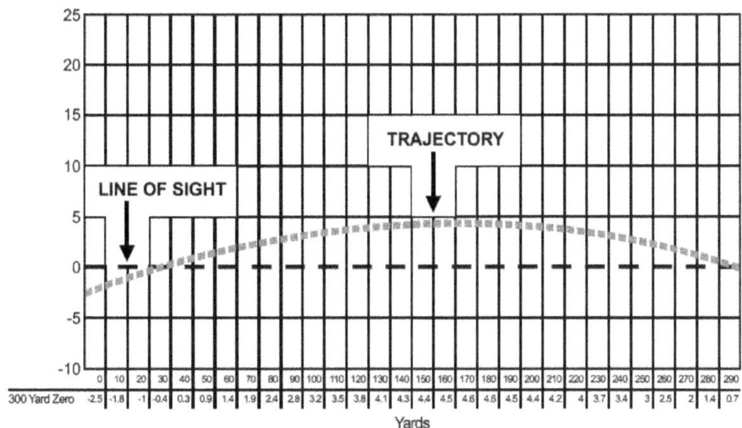

Figure 9-5. Trajectory at 0 to 300 Yards BZO Showing Rise and Fall of Bullet in Inches.

Holds for Iron Sights at 0 to 300 Meters/Yards

Within the BZO range of a Service rifle (i.e., 300 meters/yards) the aiming point is center mass and there is no offset. However, the Marine can employ offset aiming to compensate for the elevation required to engage a small target (e.g., head shot) inside the BZO of the weapon. To employ offset aiming, the point of aim (i.e., sight picture) is shifted to a predetermined location on or off the target to compensate for a known condition (e.g., distance, size of target). For a Service rifle with an established 300-meter/yard BZO, the highest point of the bullet's trajectory is 4 2/3 inches above the line of sight at 180 yards. To make an accurate head shot at this range, the Marine must hold the tip of the front sight post at the chin, rather than the T-box (see fig. 9-6 on page 9-6).

> *Note:* Holds for elevation for the RCO are made using the bullet drop compensator. This information is covered in chapters 5 and 8.

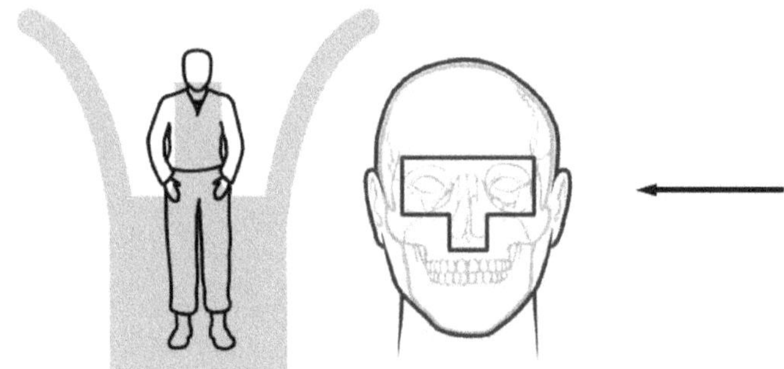

Figure 9-6. Offset Aiming Point for Head Shot at 180 Yards.

Trajectory 0 to 46 Meters/50 Yards

Concerning trajectory of the round at 0 to 46 meters/50 yards, the following applies:

- If shots are fired on a target closer than 33 meters/36 yards, they will be below sight picture. To be accurate at ranges less than 33 meters/36 yards, the Marine must aim higher than the intended impact point.
- If the bullet rises above the line of sight at 46 meters/50 yards, and the 300-yard zero will be 1-inch high; therefore, at 46 meters/50 yards the Marine must aim lower than the intended impact point.

See figure 9-7.

> *Note:* The ballistics data in figure 9-7 is calculated based on yards.

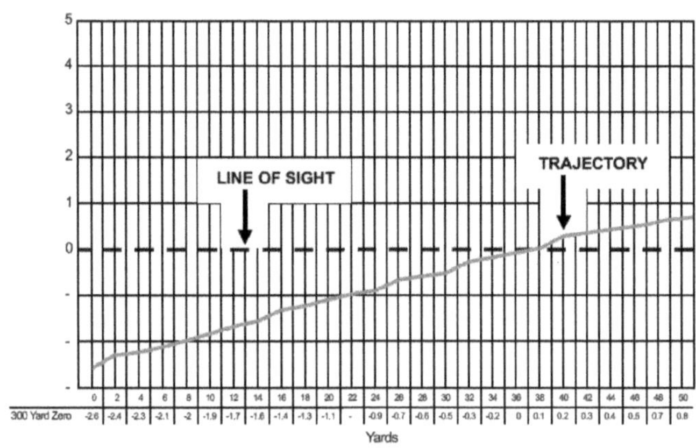

Figure 9-7. Trajectory at 0 to 50 Yards.

Trajectory at 300 to 500 Meters/Yards

The Marine employs offset aiming to compensate for the elevation required to engage a target beyond the BZO capability of the Service rifle when there is not enough time to adjust the sights. When a Service rifle is zeroed at 300 meters/yards, the bullet will cross the line of sight at 300 meters/yards and continue on its downward path of the trajectory. Figure 9-8 presents the bullet drop in inches after 300 meters/yards. Therefore, offset aiming is employed to engage the target beyond 300 meters/yards.

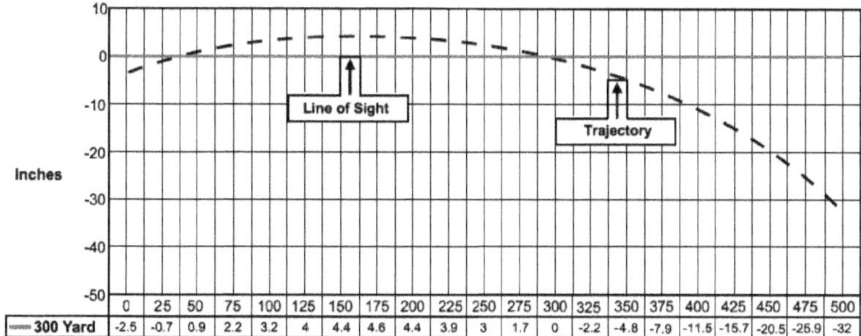

	0	25	50	75	100	125	150	175	200	225	250	275	300	325	350	375	400	425	450	475	500
300 Yard	-2.5	-0.7	0.9	2.2	3.2	4	4.4	4.6	4.4	3.9	3	1.7	0	-2.2	-4.8	-7.9	-11.5	-15.7	-20.5	-25.9	-32

Figure 9-8. Trajectory 0 to 500 Yards BZO Showing Rise and Fall of Bullet in Inches.

Holds for Iron Sights at 300 to 500 Meters/Yards

To employ offset aiming, the point of aim (i.e., sight picture) is shifted to a predetermined location on or off the target to compensate for a known condition (e.g., distance). When using offset aiming to engage a target with iron sights (i.e., the Service rifle is set on 300-meter/yard BZO) at a range estimated to be beyond 300 meters out to 400 meters, hold the tip of the front sight post at the neck (see fig. 9-9). Any distance beyond 400 meters would require holding the tip of the front sight post higher; thereby, masking the target (see fig. 9-9). A hasty sight setting should be applied at ranges beyond the Service rifle's BZO.

> ***Note:*** Holds for elevation for the RCO are made using the bullet drop compensator. This information is covered in chapters 5 and 8.

Figure 9-9. Offset Aiming Points for Elevation Outside Battlesight Zero.

Offset Aiming for Windage and Elevation

Known Strike of the Round

Known strike of the round is an offset aiming technique that shifts the aiming point (i.e., sight picture) to compensate for rounds that strike off target center. The known strike of the round technique is used if the strike of the round is known. To engage a target using this technique, the Marine aims an equal distance from center mass opposite the known strike of the round. For example, if the round strikes high and left, the Marine aims an equal and opposite distance low and right.

CHAPTER 10 Engagement Techniques

In a combat environment, targets can present themselves with little or no warning. Close-range engagements do not allow for, nor require, refined aiming techniques used in long-range precision engagements. Therefore a greater emphasis is placed on rapid presentation of the weapon. Controlled tension of major muscle groups is needed to present, stabilize, and manage recoil of the weapon. Presentation results in the sights being automatically aligned as soon as stock weld is achieved. To maintain an advantage, the Marine carries his weapon in a position that permits the Service rifle to be both easily carried and presented as quickly as possible. A carry is established based on both the sling and the situation, such as moving in a close quarters environment or moving over or under objects.

Service Rifle Presentation

From the Controlled Carry with the Three-Point Sling

The Marine uses the controlled carry with a three-point sling when no immediate threat is present as follows:

- As a target presents itself, grasp the pistol grip and bring the weapon up to place the toe of the weapon firmly in the shoulder. At the same time, bring the muzzle up by raising the support hand, allowing the Service rifle butt to pivot in the shoulder.
- As the Service rifle is being presented, take the Service rifle off **SAFE** and place the trigger finger on the trigger. At the same time, inhale, filling the lungs with air. Let your breathing assist you in acquiring sight picture and stability of hold.
- As the stock makes contact with the cheek, level the Service rifle to obtain proper stock weld. Do not move the head down to meet Service rifle stock.

> ***Note:*** If the Service rifle is in the shoulder properly, the aiming eye will be able to look through the optic/rear sight as soon as the stock makes contact with the cheek.

- As the sight(s) becomes level with the aiming eye, visually locate the target through the optic/rear sight aperture and obtain sight picture.

From the Alert or Ready Carry

The Marine uses the alert carry when adversary contact is likely or for moving in close terrain (e.g., urban or jungle environment). The ready carry is used when enemy contact is imminent.

To present the Service rifle from the alert, perform the following steps once a target appears:

- While looking at the target, bring the muzzle up by raising your support hand, allowing the Service rifle butt to pivot in your shoulder. At the same time, pull the Service rifle firmly into the pocket of your shoulder.
- As the Service rifle is being presented, take the Service rifle off **SAFE** and place your trigger finger on the trigger and inhale, filling the lungs with air.
- As the stock makes contact with your cheek, level the Service rifle to obtain a proper stock weld. Do not move your head down to meet the Service rifle stock.

> ***Note:*** If the Service rifle is in the shoulder properly, the aiming eye will be able to look through the optic/rear sight as soon as the stock makes contact with the cheek.

- As the sight(s) becomes level with the aiming eye, visually locate the target through the optic/rear sight aperture and obtain sight picture.

From the Tactical Carry

The Marine uses the tactical carry with a web sling when no immediate adversary is present. This carry permits the Service rifle to be easily carried for long periods of time, but does not permit the quickest presentation to a target. If the situation changes and an adversary presents itself, the Marine performs the following steps to present the Service rifle from the tactical carry:

- Extend the Service rifle toward the target, keeping the muzzle slightly up so that the buttstock clears all personal equipment. Continue to look at the target.
- As the Service rifle is being presented, take the Service rifle off **SAFE**, and place your trigger finger on the trigger.
- Level the Service rifle while pulling it firmly into your shoulder to obtain proper stock weld. Do not move your head down to meet the stock of the Service rifle.

> ***Note:*** If the Service rifle is in the shoulder properly, the aiming eye will be able to look through the optic/rear sight as soon as the stock makes contact with the cheek.

- As the sight(s) becomes level with the aiming eye, visually locate the target through the optic/rear sight aperture and obtain sight picture.

Shot Delivery Techniques

The Marine must maintain the ability to react instinctively in a combat environment—day or night. However, speed alone does not equate to effective target engagement. The Marine must fire only as fast as he can fire accurately. He should never exceed his physical capability to engage a target effectively. Shot delivery techniques are employed to produce accurate shots on target based on the size and distance to the target. Shot delivery techniques include two shots and failure to stop.

Two Shots

In combat, it may not always be possible to eliminate a target in a single engagement, regardless of how well the fundamentals are applied, because two shots may not cause enough trauma to the body to eliminate the target. Two aimed shots fired in rapid succession to the target increases the amount of trauma (i.e., shock, blood loss) and increases the chance of incapacitation of the target. There are two methods to execute the two-shot technique: the controlled pair and the hammer pair.

Controlled Pair

A controlled pair—

- Is two aimed shots fired upon a target in rapid succession, sight picture is acquired for both shots.
- Is an immediate target engagement technique for targets at ranges where sight picture is critical to accuracy, but distance will vary based on the individual Marine's ability.

The size and distance to the target will affect how quickly two shots can be delivered on the target. The intent is to fire two shots quickly so that the second shot is fired before the target can react to the first shot. The speed at which two shots are fired depends on the ability of the Marine and how fast he can reacquire his front sight. The speed of reacquiring sights will depend on how well recoil is managed. The better the Marine manages recoil, the faster the second shot will break. The Marine must not compromise accuracy for speed. The key to successful target engagement is to fire only as quickly as the Marine can fire effectively.

Controlled pair is the preferred technique of delivering two rapidly fired shots at ranges of greater than 15 yards. To employ a controlled pair, perform the following:

- Present the weapon to the target.
- Acquire sight picture, fire a shot, and recover the sights back on target.
- Reestablish sight picture and fire a second shot in rapid succession to the first.

Hammer Pair

A hammer pair—

- Is two shots fired in rapid succession with just one sight picture.
- Is fired at close ranges where sight picture is not as critical to accuracy and the distance will vary based on the individual Marine's ability.

To fire a hammer pair, present the weapon to the target. Once you have a sight picture on center mass of the target, fire two rapid shots without regaining sight picture. The Marine must trust his firing position and recoil management to fire the second round without reacquiring his sights. Proper body position and practice should enable the Marine to fire as fast as the trigger can be manipulated.

Failure to Stop

A failure to stop is—

- A pair fired to the torso, followed by an additional shot to an alternate aiming point (i.e., T-box in head or pelvic girdle). Time and distance to the target will determine if the initial pair is delivered via a controlled pair or a hammer pair.
- An assessment of the target following an engagement where the target is not incapacitated, followed by a single shot fired to an alternate aiming area. Assessing the situation following two shots enables the Marine to break out of the tunnel vision often associated with firing in combat, enabling him to determine follow-on action.
- A pair fired to the torso where the target still poses a threat.
- A pair fired when the torso shots have failed to stop or eliminate the target. There may be numerous reasons why body shots may not have been successful (e.g., body armor, psychological or physiological reactions to a violent encounter, ballistic failure, drugs).

T-Box

A shot in the T-box of the head is considered an incapacitating shot. The T-box is the primary alternate aiming point, because one shot to the brain has the best chance of immediately incapacitating an adversary. A frontal shot should be placed within the T-box, which is located from the brow to the bottom of the nose and from eye to eye (see fig. 10-1). A T-box shot easily penetrates the head with minimal deflection or energy loss.

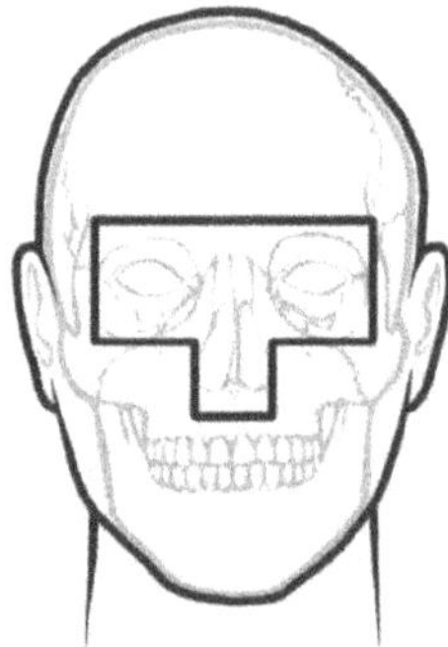

Figure 10-1. T-Box.

Terminal ballistics is what a projectile does once it strikes its target and the effects that the projectile has on a target. The goal of any engagement is immediate incapacitation of a target. The Marine Corps advocates a center mass aiming point in the center of the chest of an adversary. Center mass in the chest targets the heart and other major vascular structures.

A well-placed shot in the chest will cause the loss of massive amounts of blood very rapidly. A secondary aiming point is in the T-box of the head, which is located from the brow to the bottom of the nose and from eye to eye. Placing a round in the T-box of the head will increase the chances of incapacitating the adversary. While T-box shot will be more likely to produce immediate incapacitation, it presents a smaller target than the chest and is a more difficult shot to make.

Pelvic Girdle

A shot to the pelvic girdle is an immobilizing shot, which means that the target will go down, but not necessarily be eliminated. The pelvic girdle should only be used if there is no possible chance of engaging the T-box.

Technique

To perform a failure to stop—

- Fire two shots rapidly on a target.
- Assess the situation.
- Slow down and acquire sight picture on the alternate aiming area if the target has not been eliminated.
- Fire one single precision shot on the alternate aiming area.
- Search and assess.

Three-Round Burst

When set on **BURST**, the design of the Service rifle permits three shots to be fired from a single trigger pull. The rounds fire as fast as the weapon will function and cause the muzzle to climb during recoil. The ability to manage recoil is extremely important when firing the Service rifle on **BURST**. To achieve the desired effect (i.e., three rounds on target), the Marine must control the jump angle of the weapon to maintain the sights on target. At short ranges (i.e., 25 meters or less), firing on three-round burst can be an effective technique to place rounds on a man-sized target quickly to increase trauma on the target. To execute the three-round burst technique—

- Place the selector lever on **BURST** (see fig. 10-2).
- Aim center mass and acquire sight picture.
- Press the trigger once for the three-round burst.
- Manage recoil through controlled muscular tension in the position.

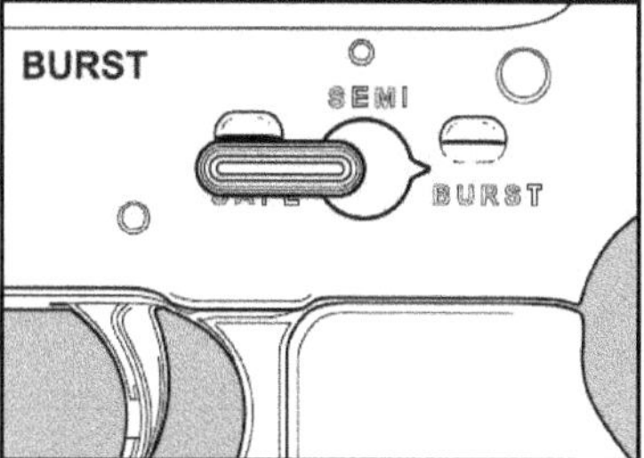

Figure 10-2. Service Rifle on Burst.

Single Precision Shot

If a target is at a long range or if the target is small (i.e., partially exposed), it can best be engaged with a single, precision shot. The Marine's stability of hold and sight alignment are more critical to accurate engagement of long range or small targets. To engage a target with the single shot technique, the Marine must slow down the application of the fundamentals and place one well-aimed shot on target.

Proper breath control is critical to the aiming process at longer ranges or when precision (e.g., small target) is necessary, because breathing causes the body to move. This movement transfers to the Service rifle, making it impossible to maintain proper sight picture. Breath control allows the Marine to fire the Service rifle at the moment of least movement.

It is critical that the Marine interrupts his breathing at a point of natural respiratory pause before firing a long-range or precision shot from any distance.

A respiratory cycle lasts approximately 4 to 5 seconds. Inhaling and exhaling each require about 2 seconds. A natural pause of 2 to 3 seconds occurs between each respiratory cycle. The pause can be extended up to 10 seconds. During the pause, breathing muscles are relaxed and the sights settle at their natural point of aim. To minimize movement, the Marine must fire the shot during the natural respiratory pause. The basic breathing technique is as follows:

- Assume a firing position.
- Stop breathing at your natural respiratory pause and make final adjustments to your natural point of aim.
- Breathe naturally until the sight picture begins to settle.
- Take a slightly deeper breath.
- Exhale and stop at the natural respiratory pause.
- Fire the shot during the natural respiratory pause.

> *Note:* If the sight picture does not sufficiently settle to allow the shot to be fired, resume normal breathing and repeat the process.

Post-Engagement Technique: Search and Assess

After the Marine engages a target, he must immediately search the area and assess the results of his engagement. Searching and assessing enables the Marine to avoid tunnel vision that can restrict the focus so that an indication of other targets may be overlooked.

Purpose of Search and Assess

The Marine searches the area for additional targets or cover and then assesses the situation to determine if he needs to—

- Reengage a target.
- Engage a new target.
- Take cover.
- Assume a more stable position.
- Cease engagement.

Technique for Search and Assess

To search and assess, the Marine performs the following steps:

- Keeps the buttstock in the shoulder and lowers the muzzle of the Service rifle slightly to look over the sights.
- Places the trigger finger straight along the receiver.
- Searches the area and assesses the situation/target by moving his head and eyes. It is not necessary to move the Service rifle with the head and eyes. Keep both eyes open to increase the field of view.

- Determines that the area is clear of enemy threat, places the Service rifle on **SAFE**, cants the weapon, and observes the chamber area to ensure that the bolt is forward.
- Determines that the fight is over. Once this determination is made, the Marine conducts a chamber check, drops the magazine to observe if adequate rounds are present, and conducts a tactical reload as necessary.

Higher Profile Searching and Assessing

Depending on the tactical situation, the Marine may choose to increase his area of observation by searching and assessing to a higher profile position.

Prone to Kneeling

After searching and assessing at the prone position, move to a kneeling position by performing the following steps:

- Lower the Service rifle butt out of your shoulder, while maintaining control of the pistol grip.
- Drop your support hand to the deck, bringing it back, and pushing up off the deck to both knees (see fig. 10-3).
- Grasp the rail system/fore grip/grip pod with your support hand and place the Service rifle butt in the pocket of your shoulder.
- Assume a kneeling position, search, and assess.

Figure 10-3. Pushing Up Off the Deck to Both Knees.

Sitting to Kneeling

After searching and assessing at the sitting position, move to a kneeling position by performing the following steps:

- Maintain control of the Service rifle with the Service rifle butt in your shoulder.
- Uncross your legs to an open-leg position.
- Tuck your firing-side foot underneath the support-side thigh as close to your buttocks as possible (see fig. 10-4). Lean forward and to the firing-side and roll onto your firing knee to a kneeling position, search, and assess (see fig. 10-5 and fig. 10-6).

Kneeling to Standing

After searching and assessing at the kneeling position, maintain control of the Service rifle with the Service rifle butt in your shoulder, standing while continuing to search and assess.

Figure 10-4. Tucking the Right Foot.

Figure 10-5. Rolling Up to a Kneeling Position.

Figure 10-6. Assuming a Kneeling Position.

CHAPTER 11

Target Detection and Range Estimation

To be proficient, the Marine rifleman must be able to detect and determine the range to targets in order to accurately engage the targets.

Target Detection and Indicators

There are many variables affecting the Marine's ability to detect and determine the range to combat targets. Enemy targets on the battlefield may be single or multiple, stationary or moving, or completely hidden from view. Success in locating an enemy target will depend upon the observer's position, his skill in searching an area, and his ability to recognize target indicators. During annual rifle range training, Marines should familiarize themselves with sight pictures at 200, 300, and 500 yards.

Most combat targets are detected at close range by smoke, flash, dust, noise, or movement; and they are usually seen only momentarily. Target indicators are anything that reveals an individual's position to the enemy. These indicators are grouped into three general areas: movement, sound, and improper camouflage.

Movement

The human eye is attracted to movement, especially sudden movement, but the Marine does not need to be looking directly at an object to notice movement. The degree of difficulty in locating moving targets depends primarily on the speed of movement. A slowly moving target will be harder to detect than one with quick jerky movements.

Sound

Sound can also be used to detect an enemy position. For example, movement, rattling equipment, or talking can make sound. Sound only provides a general location of the enemy, making it difficult to pinpoint a target by sound alone. However, sound can alert the Marine to the presence of a target and increase his probability of locating it through other indicators.

Improper Camouflage

There are three indicators caused by improper camouflage: shine, outline, and contrast with the background. Most targets on the battlefield are detected because of improper camouflage. However, sometimes an observation post or enemy firing position will blend considerably with the natural background. Only through extremely thorough and detailed searching will these positions be revealed.

Shine

Shine is created from reflective objects, such as metal or glass, pools of water, and even the natural oils from the skin. Shine acts as a beacon to the target's position.

Outline

Most enemy soldiers will camouflage themselves and their equipment and positions. The outline of objects, such as the body, head and shoulders, weapons, and web gear, are recognizable even from a distance. The human eye will often pick up a recognizable shape and concentrate on it even if the object cannot be identified immediately. The reliability of this indicator depends upon visibility and the experience of the observer.

Contrast with the Background

Indicators in this category include objects that stand out against or contrast with a background because of differences in color, surface, and shape. For example—

- A target wearing a dark uniform would be clearly visible in an area of snow or sand.
- Geometric shapes, such as helmets or rifle barrels, can be easy to detect in a wooded area.
- Fresh soil around a fighting hole contrasts with the otherwise unbroken ground surface.

While observing an area, take note of anything that looks out of place or unusual and study it in more detail. This will greatly increase the chances of spotting a hidden enemy.

Target Identification and Location

When attempting to identify and locate a target, it is important to have a good position for observation and techniques.

Position for Observation

A good position for observation is one that offers maximum visibility of the area while affording cover and concealment. The optimal observation position should allow the Marine to scan all areas of observation and offer enough concealment to prevent his position from being detected.

The Marine should avoid positions that are obvious or stand out, such as a lone tree in a field or a pile of rocks on a hill. These positions may be ideal points for easy observation, but they will also make it easier for the enemy to locate the Marine.

Methods for Searching an Area

When searching an area, the Marine will be looking for target indicators. There are two techniques for searching an area: the hasty search and the detailed search.

Hasty Search

When the Marine moves into a new area, he must quickly check for enemy activity that may pose an immediate danger. This search is known as the hasty search and should take about 30 seconds, depending on the terrain. The Marine should quickly—

- Glance at various points throughout the area rather than sweeping his eyes across the terrain in one continuous movement.
- Search the area nearest him first, since it poses the greatest potential for danger.

This hasty method of search is very effective, because it takes advantage of peripheral vision. Peripheral vision enables the detection of any movement within a wide area around the object being observed, as the eyes will be focused briefly on specific points (i.e., areas that may provide cover or concealment for the enemy).

Detailed Search

A detailed search is a systematic examination of a specific target indicator or of the entire observation area. A detailed search—

- Should be conducted immediately on target indicators that were located during the hasty search.
- Should be made from top-to-bottom or side-to-side, observing the entire object in exact detail.

If multiple indicators were observed during the hasty search, the detailed search should begin with the indicator that appears to pose the greatest danger.

After a thorough search of target indicators, or if no indicators were located during the hasty search, a detailed search should be made of the entire observation area. The 50-m overlapping strip method is normally used.

Normally, the area nearest the observer offers the greatest potential danger and should be searched first, beginning the search at one flank, systematically searching the terrain at the front in 180-degree arcs, searching everything in exacting detail, 50 meters in depth (see fig. 11-1 on page 11-4).

After reaching the opposite flank, systematically cover the area between 40 and 90 meters from your position. The second search of the terrain includes approximately 10 meters of the area examined during the first search. This technique ensures complete coverage of the area.

Continue the overlapping strip search method for as far as you can see.

Observation Methods

The combat situation will dictate the method of observation of an area. Usually, the observation method will include a combination of hasty and detailed searches.

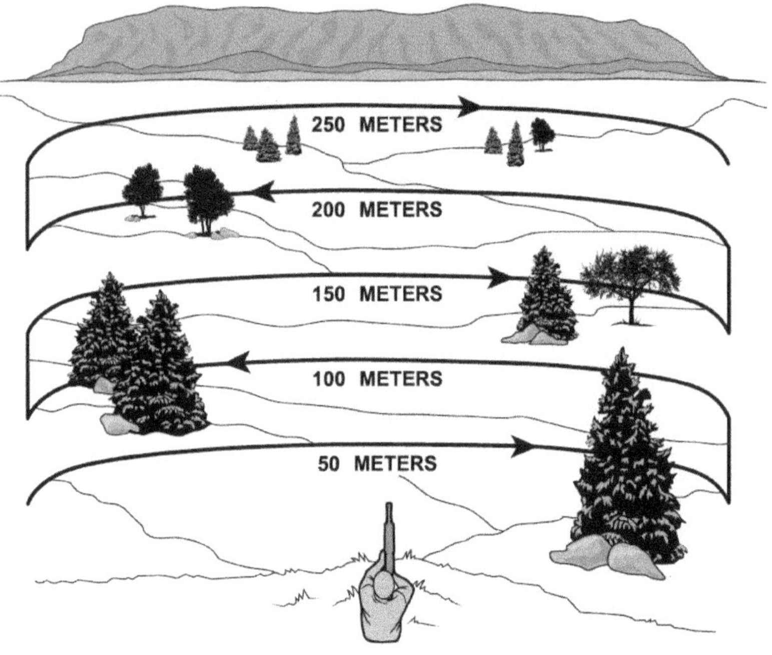

Figure 11-1. Detailed Search.

Sequence of Observation

Often, a two-man team will conduct observation. One team member should constantly observe the entire area using the hasty search technique and the other team member should conduct a detailed overlapping strip search.

If observing alone, a plan should be devised to ensure that the area of observation is completely covered. When entering a new area, immediately conduct a hasty search. Since a hasty search may fail to detect some indicators, periodically conduct a detailed search of the area. A detailed search should also be conducted any time that your attention has been diverted from the search area.

Remembering Target Location

Most targets are seen briefly and most areas contain multiple targets. Once you have located a target indicator, you will need to remember its location in order to engage it successfully. To help remember the location of a target, select a known feature and use it as a reference point to determine the distance and general direction to the target.

Range Determination and Estimation

To engage targets in a combat environment, the Marine must determine the distance from his location to a known point. This is known as range estimation. The ability to determine range is a skill that must be developed if the Marine is to successfully engage targets at unknown distances. Precise range estimation—

- Enhances accuracy.
- Enhances the chance of survival.
- Determines if a target can be effectively engaged using the Service rifle's existing BZO or if a new sight setting or point of aim is required.

Range Estimation Methods

Rifle Combat Optic Bullet Drop Compensator

The RCO has a ranging feature that can be used to estimate range to a target (see fig. 11-2). The base of the chevron and horizontal stadia lines represent 19 inches (i.e., the average width of a man's shoulders) at each of the indicated ranges. Range the target by placing the base of the chevron or a stadia line on the target to determine its range.

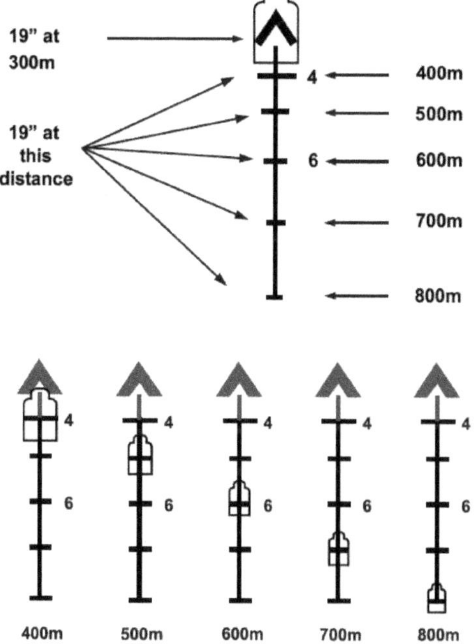

Figure 11-2. Ranging Feature on Rifle Combat Optic.

Service Rifle Front Sight Post Method

The area of the target covered by the Service rifle's front sight post can be used to estimate range to a target (see fig. 11-3). The Marine should note the appearance of the front sight post on a KD target and use this as a guide to determine range over an unknown distance. Because the apparent size of the target changes as the distance to the target changes, the amount of the target covered by the front sight post varies based on the range. Also, the Marine's eye relief and perception of the front sight post affect the amount of the target that is visible. To use this method, the Marine must apply the following guidelines:

- The front sight post covers the width of a man's chest or body at approximately 300 meters.
- If the target is less than the width of the front sight post, the target is in excess of 300 meters and the Service rifle's BZO cannot be used effectively.
- If the target is wider than the front sight post, the target is less than 300 meters and can be engaged point of aim/point of impact using the Service rifle's BZO.

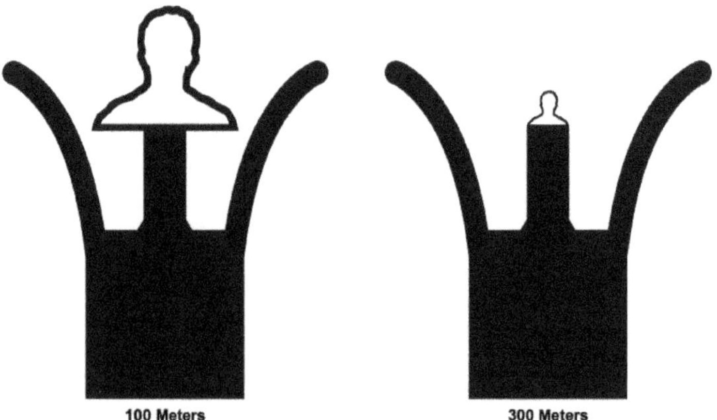

Figure 11-3. Service Rifle Front Sight Post Method.

Unit of Measure Method

To use unit of measure to determine distance, the Marine visualizes a distance of 100 meters on the ground and then estimates how many of these units will fit between him and the target. This determines the total distance to the target (see fig. 11-4).

The greatest limitation of this method is that its accuracy is related to the amount of visible terrain. For example, if a target appears at a range of 500 meters or more and only a portion of the ground between the Marine and the target can be seen, it becomes difficult to use the unit of measure method to estimate range accurately. The Marine must practice this method frequently to be proficient. Whenever possible, the Marine should select an object, estimate the range, and then verify the actual range by either pacing or using another accurate measurement.

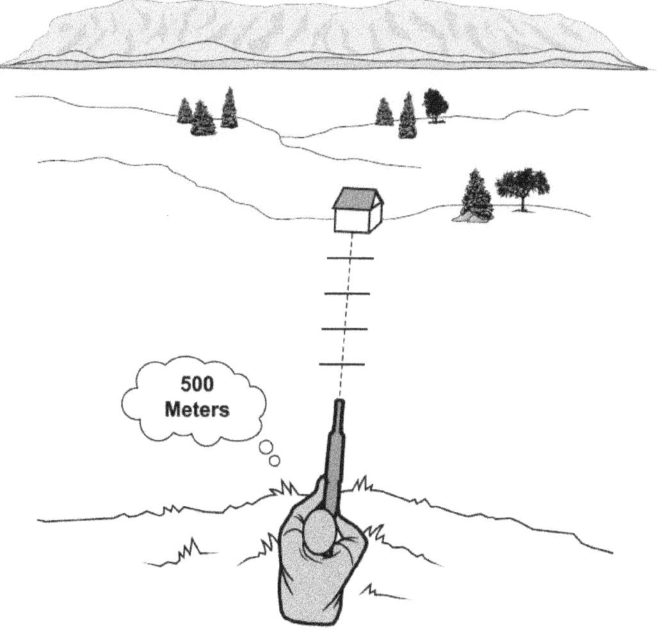

Figure 11-4. Unit of Measure Method.

Visible Detail Method

The amount of detail seen at various ranges can provide the Marine with an estimate of the target's distance. To use this method, the Marine must be familiar with the size and various details of personnel and equipment at known distances. Visibility (e.g., weather, smoke, darkness) limits the effectiveness of this method. The Marine should—

- Observe a man while he is standing, kneeling, and in the prone position at known ranges of 100 to 500 meters.
- Note the man's size, characteristics/size of his uniform and equipment, and any other pertinent details. Use these details as a guide to determine range over an unknown distance.
- Study the appearance of other familiar objects, such as rifles and vehicles.

To apply the visible detail method, the Marine should use the following general guidelines to determine distance:

- At 100 meters, the target is clearly observed in detail and facial features are distinguished.
- At 200 meters, the target is clearly observed, but there is a loss of facial detail. The color of the skin and equipment are still identifiable.
- At 300 meters, the target has a clear body outline, face color usually remains accurate, but the remaining details are blurred.

- At 400 meters, the body outline is clear, but any remaining detail is blurred.
- At 500 meters, the body shape begins to taper at the ends and the head becomes indistinct from the shoulders.
- At 600 meters, the body appears wedge-shaped and headless.

Bracketing Method

The bracketing method estimates both the shortest possible distance and the greatest possible distance to the target. For example, the Marine estimates that a target can be as close as 300 meters/yards, but it could be as far away as 500 meters/yards. The estimated distances are averaged to determine the estimated range to the target. For example, the average of 300 meters/yards and 500 meters/yards is 400 meters/yards.

Halving Method

To use the halving method, the Marine estimates the distance halfway between him and the target, then doubles that distance to determine the total distance to the target. The Marine must take care when judging the distance to the halfway point, because any error made in judging the halfway distance is doubled when estimating the total distance.

Combination Method

The methods previously discussed require optimal conditions with regard to the target, terrain, and visibility in order to obtain an accurate range estimation. Since rarely do optimal conditions exist, the Marine should select two different methods to determine range, then compare the results; or two Marines can each select a different range determination method, then compare the results. The two results are then averaged, which should result in a range that is close to the actual range to target.

Factors Affecting Range Estimation

The following specific factors will affect the accuracy of estimation. The Marine must be aware of these factors and attempt to compensate for their effects.

Nature of the Target

The following applies concerning the nature of the target:

- On a clear day, an object with a regular outline (e.g., a steel helmet, rifle, vehicle) will appear to be closer than one with an irregular outline (e.g., a camouflaged object).
- A target that contrasts with its background will appear to be closer than a target that blends in with its background.
- A partially exposed object will appear to be farther away than it is.
- A target will appear to be farther away if the target is smaller than the objects surrounding it.

Nature of the Terrain

Terrain that—

- Slopes upward gives the illusion of shorter distance.
- Slopes downward gives the illusion of greater distance.
- Contains dead space makes the target appear to be closer.
- Contains smooth material (e.g., sand, water, snow) gives the illusion of greater distance.

Limited Visibility

The position of the light source significantly affects the Marine's ability to estimate range. Other factors that affect range estimation are smoke, fog, rain, angled light, or anything that obscures the battlefield. Other factors that affect limited visibility as follows:

- When the sun is bright, a target will appear further away.
- When the sky is overcast, a target will appear closer.
- When obstacles (e.g., trees) are located between the Marine and the target, they can distract the Marine, and the target will appear farther away.
- When there is contrast on the target (e.g., color variation) between the target and the background, the target will appear closer. If there is little or no contrast between the target and the background, the target will appear farther away.

CHAPTER 12

Multiple Target Engagement Techniques

When engaging multiple targets, the Marine must prioritize each target and carefully plan his shots to ensure successful target engagement. Mental preparedness and the ability to make split-second decisions are the keys to a successful engagement of multiple targets. The proper mindset will allow the Marine to react instinctively and control the pace of the battle, rather than reacting to the adversary threat.

Threat Assessment and Prioritization

After the first target is engaged, the Marine must immediately engage the next target and continue to engage targets until they are eliminated. While engaging multiple targets, the Marine must—

- Be aware of his surroundings and not fixate on just one target.
- Prioritize the targets rapidly, establish an engagement sequence, and engage the targets.
- Maintain constant awareness and continuously search the terrain for additional targets.

By observing an adversary, the Marine can determine whether or not it presents a threat according to the rules of engagement.

Normally, the combat situation will dictate the order of multiple target engagement. Target priority is based on various factors (e.g., proximity, level of threat, opportunity), but no two situations will be the same. The level of threat for each target should be determined—from most to least threatening—so that they will be engaged in succession. The target that poses the greatest threat should be engaged first, but prioritizing targets is an ongoing process. Changes in threat level, proximity, or the target itself, can cause the Marine to revise his priorities. Therefore, the Marine must remain alert to changes in a target's threat level, proximity, and any other target opportunities as the battle progresses.

Engagement Technique

To engage multiple targets, the Marine performs the following steps:

- Engages the first target with two rounds. The recoil of the Service rifle can be used to direct the recovery of the weapon on to the next target. As the weapon is coming down in its recovery, the Marine physically brings the sights onto the next desired target.

Note: Pressure is maintained on the trigger throughout recovery, and trigger control is applied at a rate consistent with the Marine's ability to establish sight picture on the desired target.

- Engages the remaining targets in a direction that maximizes his ability to support and control the weapon, when possible (e.g., when all targets are of equal level of threat).

The preceding steps are repeated until all targets have been engaged.

Box Drill

If two shots to the torso fail to eliminate one or both of the adversaries, employ a box drill (see fig. 12-1) as follows:

- Engage the first adversary with a pair of shots to the torso. Then, while utilizing the recoil of the second shot, guide the weapon over to the next target and fire a pair of shots to that torso.
- Follow through immediately, up to the same target's head, using the recoil of the last shot to move the weapon. Pause to get a clear sight picture, and fire an incapacitating shot to the head.
- Again, using the recoil of the last shot, guide back over to the first target's head, aim in, and fire an incapacitating shot.

> *Note:* This last shot is the completion of a failure to stop drill, because you would not need to fire the last shot if your first pair to the torso had incapacitated the first target. The reason the second target is engaged with the box drill is to ensure that it will not be able to engage you while you are transitioning back to the target.

- After firing the final shot on the first target, follow through and assess the situation for further action.

Figure 12-1. Execution of a Box Drill.

Firing Position

The selection and effective use of a firing position are critical to the successful engagement of multiple targets. The Marine should make a quick observation of the terrain and select a firing position that provides good cover and concealment, as well as the flexibility to engage multiple targets. If enemy targets are widely dispersed, the selected position must provide the Marine with flexibility of movement. The more restrictive the firing position, the longer it will take the Marine to eliminate multiple targets.

Prone

The prone position limits left and right lateral movement and is not recommended for engaging short-range dispersed targets. Because the elbows are firmly placed on the ground in the prone position, upper body movement is restricted.

Sitting

Like the prone position, the sitting position only allows limited lateral movement, which makes engagement of widely dispersed multiple targets difficult. To ease engagement, pivoting on the elbow can move the forward arm, but this movement can disturb the stability of the position.

Kneeling

The kneeling position provides a wider lateral range of motion, since only one elbow is used for support. The Marine moves from one target to another by rotating at the waist to move the forward arm in the direction of the target, either right or left.

Standing

The standing position allows maximum lateral movement. Multiple targets are engaged by rotating the upper body to a position where the sights can be aligned on the desired target. If severe or radical adjustments are required to engage widely dispersed targets, the Marine moves his feet to establish a new position, rather than sacrifice maximum stability of the Service rifle. This avoids poorly placed shots that can result from an unstable position.

THIS PAGE INTENTIONALLY LEFT BLANK

CHAPTER 13

Moving Target Engagement Techniques

In combat, it is unlikely that a target will remain stationary. The enemy will move quickly from cover to cover, only exposing himself for the shortest possible time. Therefore, the Marine must quickly engage a moving target before it disappears.

Types of Moving Targets

There are two types of moving targets: steady moving and stop and go.

A steady moving target moves in a consistent manner and remains in the Marine's field of vision. An example of this target would be a walking or running man.

A stop and go target appears and disappears during its movement (i.e., presents itself for only a short period of time before reestablishing cover). This target is most vulnerable to fire at the beginning and end of its movement to new cover, because the target must gain momentum to exit its existing cover and then slow down to occupy a new position. An example of this target would be an enemy moving from cover to cover.

Leads

When a shot is fired at a moving target, the target continues to move during the time the bullet is in flight. Therefore, the Marine must aim in front of the target; otherwise, the shot will fall behind the target. This is called leading a target. Lead is the distance in advance of the target that the Service rifle sights are placed, to accurately engage the target when it is moving.

Factors Affecting Lead

Factors that affect the amount of lead are the target's range, speed, and angle of movement.

Range

Lead is determined by the Service rifle's distance to the target. When a shot is fired at a moving target, the target continues to move during the time the bullet is in flight. This time of flight could allow a target to move out of the bullet's path if the round is fired directly at the target. Time of flight will increase as range to the target increases; therefore, the lead must be increased as the distance to the target increases.

Speed

If a man is running, a greater lead is required, because the man will move a greater distance than a walking man will while the bullet is in flight.

Angle of Movement

The angle of movement across the line of sight relative to the flight of the bullet determines the type (i.e., amount) of lead.

Types of Leads

Full Lead

During a full lead, the target is moving straight across the Marine's line of sight with only one arm and half the body visible. This target requires a full lead because it will move the greatest distance across the Marine's line of sight during the flight of the bullet.

Half Lead

During a half lead, the target is moving obliquely across the Marine's line of sight at a 45-degree angle. One arm and over half of the back or chest are visible. This target requires half of a full lead because it will move half as far as a target moving directly across the Marine's line of sight during the flight of the bullet.

No Lead

During no lead, the target is moving directly toward or away from the Marine and presents a full view of both arms and the entire back or chest. No lead is required. The Marine will engage this target as a stationary target because it is not moving across his line of sight.

Leading a Moving Target

A lead is held in front of a moving target to compensate for the distance the bullet will travel while the target is moving. Table 13-1 presents the distance a target will move during the flight of the bullet at various ranges and speeds.

Table 13-1. Speed of Moving Target.

Distance Target will Move in Inches During Flight of Bullet				
	Slow Walking (2 mph)	Fast Walking (4 mph)	Jogging (6 mph)	Running (10 mph)
50 yards	2	3	5	9
100 yards	4	7	11	18
200 yards	7	14	21	35

A lead should be based on something that can be visually seen and estimated with some uniformity, such as the width of a body (i.e., approximately 12 inches). When a target is moving directly across your front, you only see the side of the target; therefore the body width examples in the following subparagraphs are considered to be approximately 12 inches. The lead will vary based on the speed and distance of the target. The aiming point of the optic reticle pattern/tip of the front sight post that is centered on the leading edge of the target is a lead of approximately 6 inches (i.e., a lead of half a body width). The following guidelines apply when establishing a lead to engage a moving target at various ranges and speeds (see fig. 13-1 on pages 13-4 and 13-5). These guidelines do not consider wind or other effects of weather.

Slow-Walking Target

A slow-walking target will be moving at approximately 2 mph, directly across the line of sight (i.e., full lead) and—

- At a range of 100 meters or less. No lead is required.
- At a range of 200 meters. Lead half a body width (i.e., sight on leading edge of target) in the direction the target is moving.

Fast-Walking Target

A fast-walking target will be moving at approximately 4 mph, moving directly across the line of sight (i.e., full lead) and—

- At a range of 100 meters or less. Lead half a body width (i.e., sight on leading edge of target) in the direction the target is moving.
- At a range of 200 meters. Lead one body width in the direction the target is moving.

Jogging Target

A jogging target will be moving approximately 6 mph, directly across the line of sight (i.e., full lead)—

- At a range of 50 meters or less. Lead half a body width (i.e., sight on leading edge of target) in the direction the target is moving.
- At a range of 100 meters. Lead one body width in the direction the target is moving.
- At a range of 200 meters. Lead two body widths in the direction the target is moving.

Running Target

A running target will be moving approximately 10 mph, directly across the line of sight (i.e., full lead) and—

- At a range of 50 meters. Lead one body width in the direction the target is moving.
- At a range of 100 meters. Lead one and a half body widths in the direction the target is moving.
- At a range of 200 meters. Lead three body widths in the direction the target is moving.

Slow Walking (2 mph)

RCO	Iron Sights	Range	Lead
		100 m or less	No lead
		200 m	Half a body width (leading edge of target)

Fast Walking (4 mph)

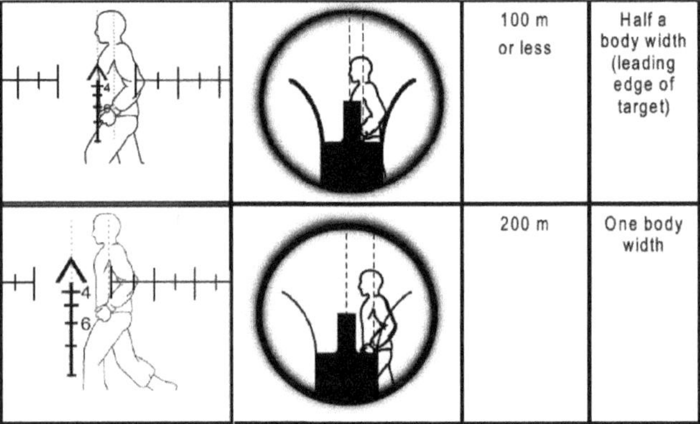

RCO	Iron Sights	Range	Lead
		100 m or less	Half a body width (leading edge of target)
		200 m	One body width

Figure 13-1. Leads for Moving Targets.

Jogging (6 mph)

RCO	Iron Sights	Range	Lead
		50 m or less	Half a body width (leading edge of target)
		100 m	1 body width
		200 m	2 body widths

Running (10 mph)

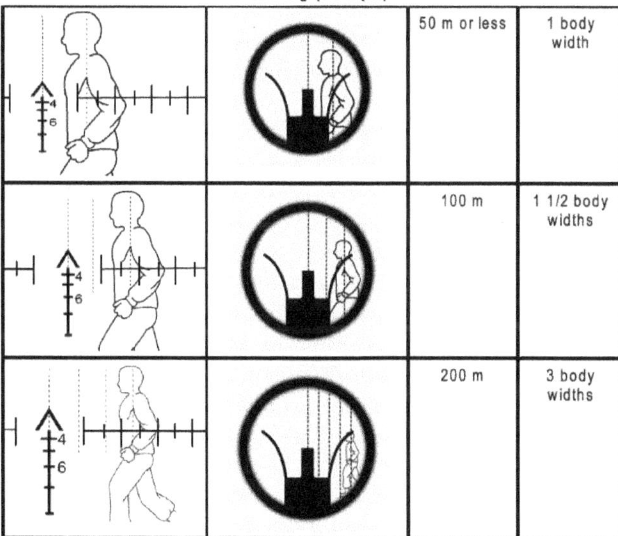

RCO	Iron Sights	Range	Lead
		50 m or less	1 body width
		100 m	1 1/2 body widths
		200 m	3 body widths

Figure 13-1. Leads for Moving Targets (Continued).

Oblique Target

An oblique target will be moving at a 45-degree angle, across the line of sight, and the lead is half of the lead that is required for a target moving directly across the line of sight.

Engagement Methods

Moving targets are the most difficult targets to engage; however, they can be engaged successfully by using either the tracking or ambush method.

The Tracking Method

The tracking method is used for a target that is moving at a steady pace over a well-determined route. If the Marine uses the tracking method, he tracks the target with the Service rifle front sight post while maintaining sight alignment and a point of aim on or ahead of (i.e., leading) the target until the shot is fired.

If employing the RCO, the Marine tracks the target with the optic's reticle pattern and the required point of aim ahead of the target until the shot is fired. When establishing a lead on a moving target, Service rifle sight(s)/optic will not be centered on the target. Instead, Service rifle sight(s)/optic will be held on a lead in front of the target (see fig. 13-2). The Marine will perform the following steps to execute the tracking methods:

- Present the Service rifle to the target.
- Swing the Service rifle muzzle through the target (i.e., from the rear of the target to the front) to the desired lead (i.e., point of aim). The point of aim can be on the target or some point in front of the target, depending upon the target's range, speed, and angle of movement.
- Track and maintain focus on the front sight post/optic reticle pattern while acquiring the desired sight picture.
- Engage the target once sight picture is acquired, while maintaining the proper lead.
- Follow through so that the lead is maintained as the bullet exits the muzzle.
- Continue to track in case a second shot needs to be fired on the target.

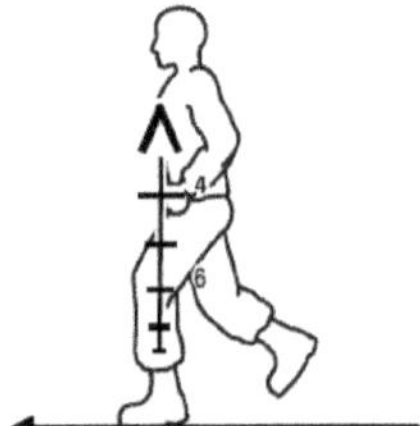

Figure 13-2. Tracking Method.

The Ambush Method

The ambush method is used when it is difficult to track the target with the Service rifle. The lead required to effectively engage the target determines the engagement point. With the sights settled, the target moves into the predetermined engagement point and creates the desired sight picture (see fig. 13-3). The trigger is pulled simultaneously with the establishment of sight

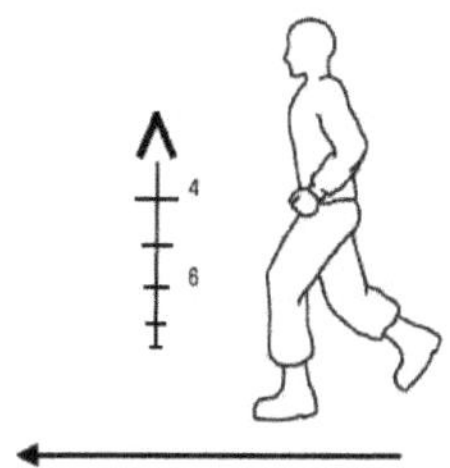

Figure 13-3. Ambush Method.

picture. To execute the ambush method, the Marine performs the following steps:

- Selects an aiming point ahead of the target.
- Obtains sight picture on the aiming point.
- Holds sight alignment/sight picture until the target moves into vision and the desired sight picture is established.
- Engages the target once sight picture is acquired.
- Follows through so that the Service rifle sight(s) are not disturbed as the bullet exits the muzzle.

A variation of the ambush method can be used when engaging a stop and go target. The Marine should look for a pattern of exposure (e.g., every 15 seconds). Once a pattern is determined, the Marine establishes a lead by aiming at a point in front of the area where the target is expected to appear, firing a shot the moment the target appears.

Modifications to Marksmanship Fundamentals

Engaging moving targets requires concentration and adherence to the fundamentals of marksmanship. The following modifications to the fundamentals of marksmanship are critical to the engagement of moving targets.

Sight Picture

Typically, sight picture is the target's center of mass. If the Marine engages a moving target, he bases his sight picture on the target's range, speed, and angle of movement. Sight picture can be established on a point of aim in front of the target.

Trigger Control

Trigger control is critical to the execution of shots with any target engagement. The Marine can apply pressure on the trigger prior to establishing sight picture, but there should be no rearward movement of the trigger until sight picture is established.

Interrupted trigger control is not recommended, because the lead will be lost or require adjustment to reassume a proper sight picture. When using the tracking method, continue tracking as trigger control is applied to ensure that the shot does not impact behind the moving target.

Follow Through

If the Marine uses the tracking method to engage moving targets, he continues to track the target during follow through so that the desired lead is maintained as the bullet exits the muzzle. Continuous tracking also enables a second shot to be fired on target if necessary.

Stable Position

A stable position steadies the Service rifle sights while tracking a moving target. To engage moving targets using the tracking method, the Service rifle must be moved smoothly and steadily with the target's movement.

> *Note:* Additional rearward pressure may be applied to the pistol grip to help steady the Service rifle during tracking and trigger control. Elbows may be moved from the support so that the target may be tracked smoothly.

CHAPTER 14

Movement

The Marine must be able to engage targets effectively in a rapidly-changing combat environment. This requires the ability to employ both lateral movement and pivoting techniques.

Lateral Movement

Lateral movement is movement in a direction other than directly toward the adversary. In the most extreme cases, the target will be offset 90 degrees or more from the direction of movement.

Firing Considerations

During lateral movement—

- Place your feet heel to toe and drop your center mass by consciously bending the knees. This will make your thighs act as shock absorbers and steady your movement to maintain the stability of your upper body, stabilizing the Service rifle sight(s) on the target. Movement should always be smooth and steady.
- Bend forward at your waist to put as much mass as possible behind your weapon for recoil management.
- Roll your foot heel to toe as you place your foot on the deck and lift it up again to provide for the smoothest motion possible.

> ***Note:*** The feet should almost fall in line during movement. This straight-line movement will keep the sights from bouncing excessively and allow a good stationary stance when needed.

- Keep your weapon at the alert or ready carry. Do not aim in on the target until you are ready to engage. You should maintain awareness of your surroundings, both to your left and right, at all times during movement.

> ***Note:*** If the Marine is moving already aimed in, he will not be aware of his surroundings. Most importantly, he will not be constantly aware of the positions of friendly forces and/or other adversaries.

Techniques

Lateral Movement to the Firing Side

An aggressive stance must be maintained throughout the entire movement and the following should be observed:

- Keep the muzzle of the weapon facing down range at the alert carry, toward the adversary.
- When moving, the placement of your feet should be heel to toe, using a combat glide. You must not overstep or cross your feet, because this can cause you to become off balance or fall.
- Keep your hips as stationary as possible. Use your upper body as a turret, twisting at your waist, maintaining proper platform with your upper body.

Lateral Movement to the Nonfiring Side

It is more difficult to engage adversaries to the nonfiring side while moving laterally. The twist required to achieve a full 90-degree offset requires proper repetitive training. The basic concept of movement must be maintained, from foot placement to platform.

Twisting at the waist will not allow the weapon to be brought to a full 90 degrees off the direction of travel, especially with nonadjustable buttstocks. The Marine will need to drop the nonfiring shoulder, rolling the upper body toward the nonfiring side. This will cause the weapon and upper body to cant at approximately a 45-degree angle, relieving some tension in the abdominal region, allowing the Marine to gain a few more degrees of offset.

Pivoting Techniques

Pivoting techniques are used to engage widely dispersed targets in the oblique and on the flanks. Pivoting skills are just as valuable in a rapidly changing combat environment as firing on the move (i.e., lateral) skills are and should only be used with the alert carry. It does not matter which way you are pivoting or which side is your strong side, as long as you always ensure that your weapon is at an exaggerated low-alert carry.

Muzzle awareness must be maintained at all times. Ensure that the muzzle does not begin to come up on target until you are facing down range (i.e., direction of adversary).

When pivoting, always look first, because that is when the target is identified. At the same time the target is being identified, you will be checking the area for tripping hazards.

When a target is located at 180 degrees behind you, execute a pivot in the direction of the adversary. To perform the pivot—

- *Look.* Turn the head and eyes to identify the target.
- *Pivot*:
  –When pivoting to the right, present the weapon as follows:
    - Turn toward the target by stepping forward with your left leg and, while pivoting, cross your left leg over your right leg.

- Continue the pivot and plant your left foot so that you are squarely facing the target and in your natural shooting stance.
- At the same time, begin presentation. Take the weapon off **SAFE** and place your trigger finger on the trigger as the weapon is presented, leveled in your shoulder, and sight picture acquired.

–When pivoting to the left, present the weapon as follows:

- Turn toward the target by stepping forward with the right leg and, while pivoting, cross your right leg over your left leg.
- Continue the pivot and plant your right foot so that you are squarely facing the target and in your natural shooting stance.
- At the same time, begin presentation. Take the weapon off **SAFE** and place your trigger finger on the trigger as the weapon is presented, leveled in your shoulder, and sight picture acquired.

- *Shoot.* Press the trigger. Ideally, the shot should be fired the instant that presentation is complete.

Forward Movement

Forward movement is movement in a direction directly toward the adversary.

Firing Considerations

During forward movement—

- Place your feet heel to toe and drop your center mass by consciously bending the knees. This will make your thighs act as shock absorbers and steady your movement to maintain the stability of your upper body, stabilizing the Service rifle sight(s) on the target. Movement should always be smooth and steady.
- Bend forward at your waist to put as much mass as possible behind your weapon for recoil management.
- Roll your foot heel to toe as you place your foot on the deck and lift it up again to provide for the smoothest motion possible.

> *Note:* The feet should almost fall in line during movement. This straight-line movement will keep the sights from bouncing excessively and allow a good stationary stance when needed.

- Keep your weapon at the alert or ready carry. Do not aim in on the target until you are ready to engage. You should maintain awareness of your surroundings, both to your left and right, at all times during movement.

> *Note:* If the Marine is moving already aimed in, he will not be aware of his surroundings. Most importantly, he will not be constantly aware of the positions of friendly forces and/or other adversaries.

Technique

An aggressive stance must be maintained throughout the entire movement and the following should be observed:

- Keep the muzzle of the weapon facing down range at the alert carry, toward the adversary.
- When moving, the placement of your feet should be heel to toe—measuring approximately 12 to 15 inches using a combat glide. You must not overstep or cross your feet, because this can cause you to become off balance or fall.
- Keep your hips as stationary as possible. Use your upper body as a turret, twisting at your waist, maintaining proper platform with your upper body.

Rearward Movement

Movement to the rear is one or two steps that will open the distance or reposition the feet. To perform rearward movement, place the feet in a toe to heel manner and drop the center body mass by consciously bending the knees utilizing a reverse combat glide.

CHAPTER 15

Low-Light Engagement Techniques

Combat targets are frequently engaged during periods of darkness or under low-light conditions. Although basic marksmanship fundamentals do not change, the principles of night vision must be applied and target detection is applied differently. In addition, during periods of darkness or low light, the Marine's vision is extremely limited.

Night Vision

The Marine—

- Must apply the techniques of night observation in order to detect potential targets.
- Must also develop skills that allow him to engage targets under these conditions.
- Can improve his ability to see during periods of darkness or low light by obtaining and maintaining night vision. Since adapting to night vision is a slow and gradual process, steps should be taken to protect night vision once it is obtained.

How to Obtain Night Vision

There are two methods used to obtain night vision.

The first method is to remain in an area of darkness for approximately 30 minutes. This area can be indoors or outdoors. The major disadvantage of this approach is that an individual is not able to perform any other tasks while acquiring night vision in total darkness.

The second method is to remain in a darkened area under low-intensity red light (i.e., light that is similar to a photographer's darkroom) for approximately 20 minutes, followed by 10 minutes in darkness without the red light. This method produces almost complete night vision adaptation, while permitting the performance of other tasks during the adjustment period.

How to Maintain Night Vision

Because the eyes take a long time to adjust to darkness, it is important to protect night vision once it is acquired. To maintain night vision—

- Avoid looking at any bright light. Bright light will eliminate night vision and require readaptation.
- Shield eyes from parachute flares, spotlights, or headlights.

- When using a flashlight to read a map or any other written material—
  - Put one hand over the glass to limit the area illuminated and the intensity of the light. Keeping one eye shut will reduce the amount of night vision lost.
  - Cover the light with a red filter to help reduce the loss of night vision.
  - Minimize the time spent using a flashlight.

How Night Vision is Affected

Some physical factors can affect night vision and reduce the ability to see clearly in low light or darkness. These factors include the following:

- Fatigue.
- Lack of oxygen.
- Long exposure to sunlight.
- Heavy smoking.
- Drugs.
- Headaches.
- Illness.
- Consumption of alcohol within the past 48 hours.
- Improper diet.

Searching Methods

Once night vision has been acquired, the Marine can locate targets by using off-center vision to scan.

> *Note:* Some daylight observation techniques (e.g., searching for target indicators) also apply during periods of darkness or low light.

Off-center vision is the technique of keeping attention focused on an object without looking at it directly (see fig. 15-1). To search for targets using off-center vision, never look directly at the object you are observing. Look slightly to the left, right, above, or below the object. Experiment and practice to find the best off-center angle for you. For most people, it is approximately 6 to 10 degrees away from the object or about a fist's width at arm's length.

> *Note:* Staring at a stationary object in the dark can make it appear to be moving. This occurs because the eye does not have a reference to determine the exact position of the object. This illusion can be prevented by visually aligning the object against something else, such as a finger at arm's length.

When scanning, the use of off-center vision to observe an area or object involves moving the eyes in a series of separate movements across the objective area.

Figure 15-1. Off-Center Vision.

A common method of scanning is to move the eyes in a figure-eight pattern (see fig. 15-2). To employ the figure-eight pattern—

- Move the eyes in short, abrupt, irregular movements over and around the area. Once a target indicator is detected, the focus is concentrated in that area, but not at it directly.
- Pause a few seconds at each point of observation, because the eyes cannot focus on a still object while in motion. The Marine should rest the eyes frequently when scanning.

> *Note:* While observing, there may be periodic blackouts of night vision because of simple fatigue. This is normal and not a cause for alarm. Night vision will return quickly once the Marine moves or blinks the eyes.

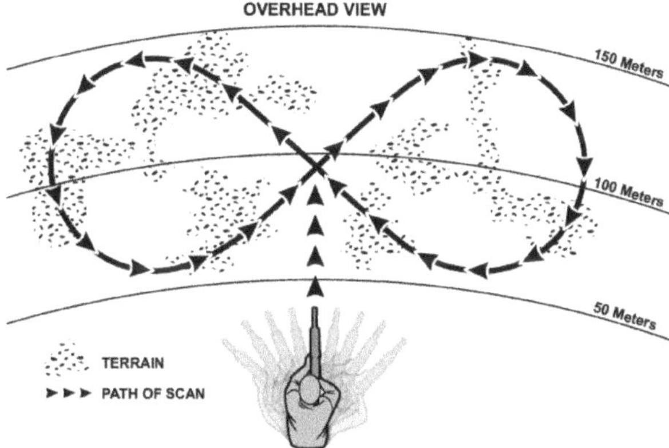

Figure 15-2. Figure Eight Scan.

It is more effective to scan from a prone position or from a position that is closer to the ground than the object being observed. This creates a silhouetted view of the object. When scanning an area, look and listen for the same target indicators as in daylight: movement, sound, and improper camouflage. Objects in bright moonlight/starlight cast the same shadows as in sunlight and sound seems louder at night than during daylight.

Illumination

Types of Illumination

There are two types of illumination that assist engagement during low light or darkness: ambient light and artificial illumination. Both ambient light and artificial illumination can affect target perception (i.e., distance and size) and night vision capabilities.

Ambient Light

The light produced by natural means (i.e., the sun, moon, and stars) is ambient light. Variations that occur in ambient light are affected by the time of day, time of year, weather conditions, terrain, and vegetation.

Artificial Illumination

The light produced by a process other than natural means is artificial light. Artificial light can be used to illuminate an area for target detection or to illuminate a specific target to pinpoint its position. The two types of artificial illumination used in combat are air and ground.

Effects of Illumination

In some combat situations, ambient light and artificial illumination may assist the Marine in locating targets. However, this light can also affect perception of the target and disrupt night vision.

The introduction of artificial light requires the eyes to make a sudden, drastic adjustment to the amount of light received. This can cause a temporary blinding because night vision was abruptly interrupted. Ambient light may also cause a blinding effect. For example, the Marine may experience temporary blindness or reduced night vision if a bright moon suddenly appears from behind the clouds. Other conditions that may affect night vision are the following:

- Light behind the Marine or light between the Marine and a target illuminates the front of the target and makes it appear closer than it is.
- Light beyond the target displays the target in silhouette and makes it appear farther away than it is. If the target is silhouetted, it is easier to see and engage.
- Air illumination devices are in constant motion as they descend to the ground. This movement creates changing shadows on any illuminated target, causing a stationary target to appear as if it is moving.

Acquiring Targets at Night

When acquiring targets at night—

- Hold your head high so that your eyes are well above your weapons sights. This will increase your field of view and improve the sharpness of detail.
- Keep both eyes open to get maximum visual coverage of the target area and improve depth perception.

To obtain sight alignment/sight picture during low light or reduced visibility—

- Flip the Service rifle's large rear sight aperture (0-2 sight) up. Using this larger aperture enables you to take greater advantage of whatever illumination is available to acquire sight alignment/sight picture.
- Obtain stock weld and try to obtain sight alignment and sight picture. There is normally enough ambient light to enable you to perceive an object as far as away 50 meters, especially if it is moving.

> ***Note:*** When Service rifle sights are placed on a dark background, such as a camouflage target, the Marine may not be able to acquire and align sights clearly. The Marine may need to rely on presentation to get the weapon on target. To check sight alignment and/or acquire the sights, point the Service rifle toward an area that provides a good contrast (e.g., the skyline), then bring the sights back on line with the target.

Artificial illumination, particularly air devices, may make the target appear to move, disrupting the ability to obtain a proper sight picture. Under this condition, the Marine may need to obtain sight alignment by focusing the sights on the lower portion of the target. This area of the target will be less affected by the shadows created by the illumination and provide a more stable aiming point. Once sight alignment has been established on this area of the target, raise the Service rifle sights to center mass and engage.

Night Aiming Devices

AN/PEQ-15

Specifications

The AN/PEQ-15 advanced target pointer illuminator aiming light is a multifunction laser device that emits visible or infrared (IR) light for precise weapon aiming and target/area illumination as follows:

- The visible aim laser provides the ability for active target acquisition in low light and close quarters combat situations without night vision devices.
- The IR aim and illumination lasers provide the ability for active, covert target acquisition in low light or complete darkness when used in conjunction with night vision devices.

The AN/PEQ-15 can be used as either a handheld illuminator/pointer or be mounted to weapons equipped with a rail system. Figure 15-3 presents the features and controls of the AN/PEQ-15.

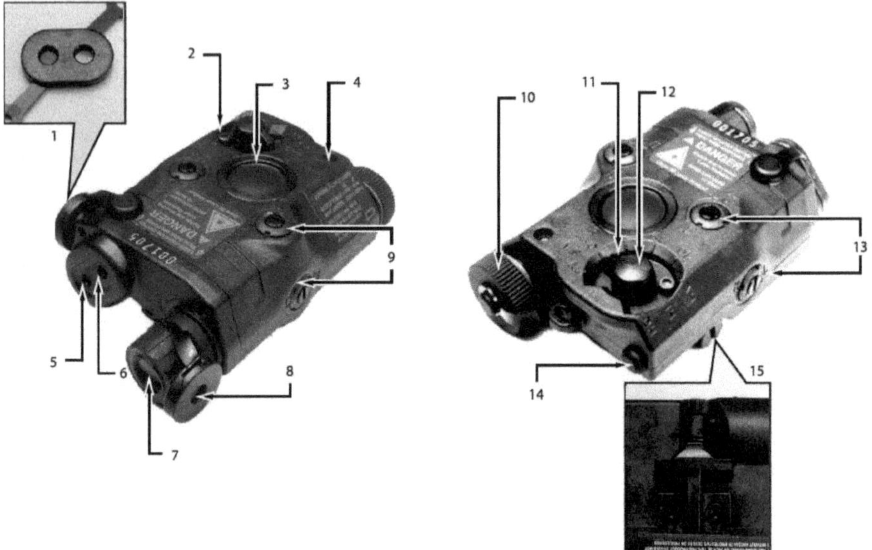

Legend:

1 Aim neutral density/opaque lens cap
2 Safety screw
3 Activation button
4 Safety screw storage location
5 Visible aim laser
6 Infrared aim laser
7 Infrared illuminator
8 Illuminator diffuser lens cap
9 Illuminator adjusters
10 Battery cap
11 LED [light emitting diode] status indicator
12 Mode selector
13 Aim laser adjusters
14 Remote jack/jack plug
15 Integral rail grabber bracket

Figure 15-3. AN/PEQ-15 Features and Controls.

Weapons Mounting

The AN/PEQ-15 can be mounted on the top (see figs. 15-4 and 15-5), left, or right side of the Service rifle. If mounted on the side of the Service rifle, the AN/PEQ-15 should be mounted on the outboard side (i.e., right-side mount for right-handed Marine) to ensure that the sling or gear does not interfere with its operation.

To mount the AN/PEQ-15:

- Loosen the clamping knob on the integral rail grabber bracket until the jaws have sufficient space to fit over the rail system.
- Position the integral rail grabber bracket on the rail, ensuring that the recoil lug is seated in the desired recoil groove of the rail. Turn the clamping knob clockwise to tighten.

Figure 15-4. Top Mount on M4/M4A1.

Figure 15-5. Top Mount on M16A4.

Note: The AN/PEQ-15 can be placed at any position (i.e., forward or aft) on the rail that is most convenient for the operator. However, if the AN/PEQ-15 is removed from the rail, the operator must make note of the position where it was zeroed and return it to that same position in order to ensure that zero is retained. Failure to properly secure and tighten the AN/PEQ-15 to the rail can lead to boresight repeatability and zeroing issues.

Boresighting and Zeroing Procedures

Boresighting. The AN/PEQ-15 is equipped with boresight adjusters for independent adjustment of the aiming and illumination beams in both elevation and azimuth. The AN/PEQ-15 incorporates a unique zero preset feature that enables the coaligned visible and IR aim lasers to be close to zeroed when initially attached to the weapon (i.e., within 4 inches vertically and horizontally of the mechanical axis of the weapon's barrel at 25 meters). To establish this zero preset—

- Rotate both the azimuth and elevation aim laser adjusters to the full counterclockwise end of travel.
- Rotate both the azimuth and elevation aim laser adjusters back two and one-half turns to align the slotted head in a 12 o'clock/6 o'clock orientation.

After establishing the zero preset or boresighting the AN/PEQ-15 weapon combination, the AN/PEQ-15 can be zeroed to the weapon via live fire on a 25-m range.

Zeroing. The following zeroing procedures should be observed:

- On a 25-m M16 zeroing target, mark the designated strike point and designated strike zone for the weapon being used (see table 15-1).

Table 15-1. AN/PEQ-15 Target Offset.

Mount	10-m Boresight Target Offset Squares	25-m M16 Target Zero Offset Squares
Top rail	VIS 2.0 right / 1.5 up IR 1.0 right / 2.5 up	VIS 2.5 left / 1.5 up IR 1.5 left / 0.5 up
Left rail	VIS 3.0 left / 0.0 IR 4.0 left / 1.0 down	VIS 2.5 right / 3.5 up IR 3.5 right / 4.5 up
Right rail	VIS 3.0 right / 4.5 down IR 4.0 right / 3.5 down	VIS 3.0 left / 7.0 up IR 4.0 left / 6.0 up

- Mount the target on an E silhouette or other suitable surface at 25 meters.
- Mount the AN/PEQ-15 to the weapon.
- Rotate the mode selector to **VIS-AL** [visible aim].
- Activate the visible aim laser in continuous mode by double-tapping the activation button.
- Direct the visible aim laser at the center of the target.
- Fire a three-round shot group and note the center of the shot group relative to the designated strike point.
- Retighten integral rail grabber bracket.
- Rotate the aim laser adjusters to move the center of the shot group to the designated strike point.
- Fire another three-round shot group and again observe the center of the new shot group relative to the designated strike point. When five out of six consecutive rounds are in the designated strike zone, the AN/PEQ-15 weapon combination is zeroed.
- Once the AN/PEQ-15 weapon combination is zeroed, apply a positive load to each adjuster by turning each one eight clicks (one-quarter turn) clockwise, then back counterclockwise to the zero position.
- Once the aiming beams are zeroed, rotate the mode selector to the **DL** [dual low] or **DH** [dual high] position to observe both the IR aiming and illumination beams. Rotate the illuminator adjusters to center the illumination beam over the IR aiming beam.

Note: The field expedient zeroing method using cowitness procedures is used when employing the RCO. It allows a quick boresight at 100 meters. Point of impact will be approximately 4 to 5 inches low at 200 meters. It can only be used with a weapon/ RCO that has an established zero. This method is performed in low light or dark conditions in which the visible laser may be seen.

- Place a target (e.g., E silhouette) 100 meters from the firing line.
- Hang a chemlight on the target at a height that can be engaged from a prone position.
- Assume a supported prone position.

- Rotate the AN/PEQ-15 mode selector to **VIS-AL**.
- Aim in on the chemlight using the tip of the RCO chevron as the aiming point.
- Activate the visible aim laser in continuous mode by double tapping the activation button.
- While the Marine with the weapon maintains a steady aim on the chemlight, the other Marine (i.e., the cowitness) finds the beam and adjusts it onto the chemlight using the AN/PEQ-15 adjusters.

The AN/PEQ-15 is considered boresighted when the tip of the chevron of the RCO is centered on the visible laser at the aiming point on the target.

AN/PEQ-16

Specifications

The AN/PEQ-16A (mini-integrated pointer illuminator module) is a multifunction laser device that emits visible or IR light for precise weapon aiming and target and/or area illumination. It is equipped with a white light illuminator. It is hand held, weapon mounted, and battery operated (requires two 3-volt batteries). See figure 15-6.

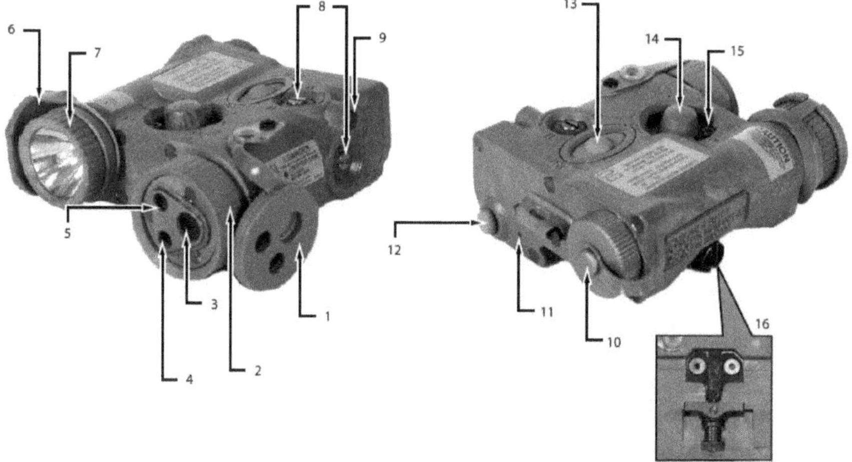

Legend:

1	Trifunction lens cap	9	Safety screw storage location
2	IR illuminator focus knob	10	Battery cap/battery compartment
3	IR illuminator	11	LED status indicator
4	Visible aim laser	12	Remote jack/jack plug
5	IR aim laser	13	Activation button
6	White light lens cap	14	Mode selector
7	White light illuminator/focus knob	15	Safety screw
8	Boresight adjusters	16	Rail grabber bracket

Figure 15-6. AN/PEQ-16.

Visible Aiming Laser. The visible aiming laser provides for active target acquisition in low light and close quarters combat situations without the need for night vision devices. It is used to provide a precision aim point or to positively identify targets at close range during the day or night, without the need of night vision devices.

Infrared Aiming Laser. The IR aiming laser is used with night vision devices to provide a precision aim point or to mark targets.

Infrared Illuminator. The IR illuminator provides for active, covert target acquisition in low light or complete darkness when used in conjunction with night vision devices. It also provides variable focused IR illumination of the intended target area. A focus knob is used to vary the IR illumination beam spread from flood to spot, based on the range and size of the area to be illuminated.

Light-Emitting Diode Status Indicator. Whenever the AN/PEQ-16 is activated, the green, light-emitting diode will light. It is located next to the battery cap.

Trifunction Lens Cap. A trifunction lens cap can be installed over the trilaser assembly. It uses an illuminator diffuser to spread the laser energy from the IR illuminator over an angle of approximately 180 degrees, allowing for illumination of a 20 ft x 30 ft x 8 ft room.

White-Light Illuminator. The white-light illuminator provides for target identification/illumination without using night vision devices. A focus knob provides a variable focus, white-light beam designed to allow for facial recognition at 25 meters. A lens cap is an opaque filter that, when installed over the white-light illuminator, significantly reduces, but does not prevent, inadvertent emission of white-light energy.

Laser Classification

The AN/PEQ-16 is a laser; therefore, all laser safety precautions should be taken in training environments. The AN/PEQ-16 operates on three different laser classifications: Class 1, Class 3A, and Class 3B. Eye damage can occur if—

- Pointed at a naked eye less than 220 meters away.
- Pointed at a person looking through binoculars less than 1300 meters away.

Weapons Mounting

The AN/PEQ-16 is equipped with an integral rail-grabber bracket that is designed for direct attachment to weapons with a rail system. The AN/PEQ-16 can be mounted on the top (see fig. 15-7), left, or right of the Service rifle. If mounted on the side of the Service rifle, the AN/PEQ-16 should be mounted on the outboard side (i.e., right-side mount for right-handed Marine) so that the sling or gear does not interfere with its operation. To mount the AN/PEQ-16 to the rail—

- Loosen the clamping knob on the integral rail grabber bracket until the jaws have sufficient space to fit over the M1913 rail.
- Position the rail grabber bracket in the desired recoil groove of the rail, pushing down and forward to ensure that the laser system is properly seated.

Figure 15-7. Top Mounted on M16A4.

- Turn the clamping knob clockwise until it is as finger tight as possible. Insert the appropriate tool (e.g., screwdriver, multipurpose tool) into the screw slot in the clamping knob and turn the clamping knob an additional three-quarter turn to properly secure the AN/PEQ-16 to the rail system.

Boresighting and Zeroing Procedures

The AN/PEQ-16 aim lasers are coaligned; therefore, a single set of adjusters moves both aiming beams. Boresighting and/or zeroing can be accomplished using either the visible or IR aim laser. Table 15-2 indicates the direction of adjuster rotation and resultant shot-group movement based on where the AN/PEQ-16 is mounted on the weapon.

Table 15-2. Mounting Configurations and Weapons Offsets.

Mount	Adjuster	Rotation	Shot Group Movement
Top	Top adjuster elevation	CW CCW	Up Down
	Side adjuster windage	CW CCW	Left Right
Left side	Side adjuster windage	CW CCW	Left Right
	Bottom adjuster elevation	CW CCW	Down Up
Right side	Top adjuster elevation	CW CCW	Up Down
	Side adjuster windage	CW CCW	Right Left
Legend: CW clockwise CCW counterclockwise			

Boresighting Procedures. The AN/PEQ-16 incorporates a unique zero preset feature that enables the coaligned lasers to be close to zeroed when initially attached to the weapon (i.e., within 4 inches vertically and horizontally of the mechanical axis of the weapon's barrel at 25 meters).

CAUTION

Do not force the adjusters beyond their end of travel. To establish this zero preset, rotate the boresight adjusters to the full counterclockwise end of travel, then rotate them back two and one-half turns. The AN/PEQ-16 can also be boresighted to the host weapon using a laser borelight system. Table 15-2 provides 10-m target offsets for this purpose. Once the AN/PEQ-16/weapon combination is boresighted, place a positive load on each adjuster by turning each one eight clicks (i.e., one-quarter turn) clockwise, then back counterclockwise to the boresight position. After establishing the zero preset or boresighting the AN/PEQ-16/weapon combination, the AN/PEQ-16 can be zeroed to the weapon via live fire at a 25-m range. Table 15-3 provides target offsets that must be applied to the 25-m M16A2/A4 zeroing target.

Table 15-3. Mounting Configuration and Weapons Offsets.

Mount	10-m Boresight Target Offset Squares	25-m M16 Target Zero Offset Squares
Top rail	VIS 1.8 left / 1.5 up IR 1.8 left / 2.6 up	IR 2.0 right / 1.0 up
Left rail	VIS 2.0 Left / 2.7 Down IR 3.4 Left / 2.7 Down	IR 3.5 right / 5.0 up
Right rail	VIS 3.4 right / 0.7 up IR 4.7 right / 0.7 up	IR 4.2 left / 3.0 up
Legend: VIS visible (as in visible aim laser)		

Zeroing Procedures. Zeroing the AN/PEQ-16 is conducted at 25 meters on an M16 25-m zeroing target by performing the following steps:

Note: AN/PEQ-16 boresight adjusters move the aiming beams at the rate of 0.2 miliradian per click. Two clicks = one box on a standard M16 25-m zeroing target.

- On a 25-m M16 zeroing target, mark the designated strike point and designated strike zone for the weapon you are using (see table 15-2).
- Mount the target on an E silhouette or other suitable surface at 25 meters.
- Mount the AN/PEQ-16 to the weapon.
- Rotate the mode selector to the **AL** [IR AIM LOW] position.
- Activate the IR aim laser in continuous mode by double tapping the activation button.
- Direct the IR aim laser at the center of the target with the use of a night vision device.
- Fire a three-round shot group and note the center of the shot group relative to the designated strike point.
- Retighten the integral rail grabber bracket.

- Rotate the boresight adjusters to move the center of the shot group to the designated strike point.
- Fire another three-round shot group and observe the center of the new shot group relative to the designated strike point.
- When five out of six consecutive rounds are in the designated strike zone, the AN/PEQ-16 weapon combination is zeroed.
- Once the AN/PEQ-16 weapon combination is zeroed, place a positive load on each adjuster by turning each one eight clicks (i.e., one-quarter turn) clockwise, then back counterclockwise to the zero position.

After mounting and zeroing the AN/PEQ-16—

- It should not be removed.
- It is effective out to 250 meters if zeroed properly. The exact range will depend on the quality of night viewing system being used.

When employing the aiming device—

- The night vision should remain on.
- The laser mode selector should be turned to the appropriate setting.
- The Marine should be ready to activate the laser when needed.

> ***Note:*** Activate the laser when presenting the weapon to a target. Activating the laser prematurely or excessively can result in the Marine's position being detected by the enemy. Activate the illuminator to check for targets or scan the area.

THIS PAGE INTENTIONALLY LEFT BLANK

APPENDIX A

Training

Training forms the foundation for building combat marksmanship skills. The content in this appendix supports training on a KD range, where Marines learn and practice basic marksmanship fundamentals and positions.

Training starts on a basic level and progresses to more combat-realistic training situations and conditions. A sound marksmanship program consists of lecture, demonstration, practical application in the form of dry and live-fire training and simulation, and evaluation and mastery before progressing to additional combat-related marksmanship skills. Training is designed as a progression of skills that builds upon previously learned skills.

Marine Corps Order 3574.2_, *Marine Corps Combat Marksmanship Programs*, dictates the requirements of the Marine Corps Service rifle training program, while detailed instructor guides provide the instructional content of the training program. The Marksmanship Program Management Section, Weapons Training Battalion, Quantico, VA, maintains the Service rifle doctrinal training program (to include Marine Corps Order 3574.2_, programs of instruction, instructor guides, and this publication.

Range Operations

The KD ranges are where Marines learn and apply marksmanship skills on a live-fire range. These ranges—

- Do not simulate combat firing.
- Provide Marines with the opportunity to learn, practice, and master marksmanship fundamentals.

Coaches teach Marines to apply marksmanship fundamentals and to analyze their shooting performance. This feedback is used to improve the application of the fundamentals of marksmanship, firing positions, and techniques of fire.

Known distance firing instills the fundamentals that are the basis for all marksmanship training. The skills that are taught during KD firing will be applied in a combat environment; therefore, specific considerations are given to firing positions, effects of weather, and zeroing. Specific requirements must be met to ensure a fair evaluation of all Marines.

Range operations effectiveness depends upon the experience and quality of its personnel. When filling specific billets, the experience level, teaching ability, communication skills, motivation, and professionalism of personnel should be considered. Each billet in the range organization has specific areas of responsibility, detailed in the following subparagraphs.

Range Officer in Charge

The range officer in charge is responsible for the overall conduct of range operations. Responsibilities of this position include the following:

- Supervising range and firing personnel.
- Coordinating all training details and relay assignments.
- Supervising placement of range flags, warning signs, and signals.
- Ensuring that range and safety regulations are followed and enforced.
- Conducting record firing and ensuring that all rules governing record firing are observed.
- Making final decisions on awarding alibis if they are contested.
- Ordering and maintaining range supplies.

Line Staff Noncommissioned Officer

The line staff noncommissioned officer (SNCO) assists the range officer in the operation of the range, including the selection of coaches or other operating personnel, enforcing range safety regulations, and monitoring the conduct of fire. This billet is usually filled with the senior SNCO assigned to the range. Responsibilities of this position include the following:

- Organizing and conducting live-fire training on ranges.
- Assigning Marines to targets, relays, and pit details.
- Assigning and supervising coaches and block noncommissioned officers (NCOs).
- Ensuring that scorekeeping procedures are followed.
- Enforcing regulations for the conduct of fire.
- Observing the firing line to ensure that all firing is conducted in a safe manner.
- Ensuring that the range box stored on the range includes hearing protection, extra databooks, pens or pencils, rags, CLP, and paint pens for marking Service rifle sights.

Tower Noncommissioned Officer

The tower NCO issues all line and firing commands. This individual should be alert to safety violations, knowledgeable of all courses of fire and associated commands, and possess good communication skills. Responsibilities of this position include the following:

- Establishing and maintaining consistency in range operations.
- Ensuring line to pit communication is maintained.
- Controlling movement of firing personnel on the range.
- Monitoring and responding to signals from the firing line.

- Monitoring pit services and reporting problems with pit service to the pit SNCO. The tower NCO serves as communication link between the pit SNCO and the block NCO.
- Maintaining the tower log to document alibis and saved rounds, by relay and target number.
- Keeping time to back up the official time kept in the pits.
- Conducting range safety brief.

Block Noncommissioned Officer

The block NCO is the primary link between the coach and the tower NCO. All communication from the coach to the line SNCO should go through the block NCO. Responsibilities of this position include the following:

- Supervising conduct of training and enforcing range and safety regulations.
- Making decisions on alibis as required and signaling the tower NCO to report the alibi.
- Monitoring pit service and reporting problems to the tower NCO.
- Authorizing movement of Marines off the firing line during slow fire stages.
- Conducting refresher training for coaches as required.

Pit Staff Noncommissioned Officer

The pit SNCO is responsible to the range officer for pit operations. He oversees and controls all pit operations and enforces pit regulations. Responsibilities of this position include the following:

- Giving pit commands during live-fire training.
- Monitoring and controlling all information received or passed over the communications system between the pits and the firing line.
- Supervising the conduct of personnel assigned to the pits.
- Ensuring that pit scorekeeping procedures are followed.
- Enforcing pit operation and safety regulations.
- Conducting the pit safety brief.
- Supervising pit verifiers in the conduct of their duties.
- Keeping the official time for all timed events.
- Training and supervising an assistant pit NCO.
- Ensuring the quality control of target preparation.

Assistant Pit Staff Noncommissioned Officer

The assistant pit NCO assists the pit SNCO in the performance of his duties. Responsibilities of this position include the following:

- Maintaining supplies in operation of the pits.
- Running working parties in the pits. This can include target factory operations and pit maintenance.

Combat Marksmanship Trainer

The combat marksmanship trainer is a subject matter expert for combat marksmanship training within his unit. Responsibilities of this position include the following:

- Assisting unit commanders in conducting the Marine Corps Combat Marksmanship Program.
- Conducting all classroom instruction during preparatory and preliminary training.
- Training and supervising combat marksmanship coaches (CMCs).

Combat Marksmanship Coach

The CMC is the individual who interacts most directly with the Marine during live-fire training. Responsibilities of this include the following:

Before Marines Arrive

The following are responsibilities of the CMC before the Marines arrive for training:

- Receives special instructions from the range safety officer or line SNCO (e.g., assists with ammunition distribution, runs up flags, positions range safety markers, prepares targets).
- Receives target assignments and course of fire.
- Discusses and evaluates weather conditions with other coaches.

> ***Note:*** If the CMC is taking over the targets of another coach, that coach should provide the CMC with the Marine's names and a quick evaluation of each one.

After Reporting to Target Assignments

The following are responsibilities of the CMC once the Marines arrive and prior to the day's firing:

- Ensure that weapons handling procedures are followed and the four safety rules are enforced throughout training.
- Muster Marines in accordance with local procedures.
- Conduct the coach's brief to include the following:
  - –Determining the course of fire.
  - –Determining the firing procedure.
  - –Reviewing the weather conditions.
  - –Filling out the databook.
  - –Reviewing fundamentals of marksmanship.
  - –Reviewing techniques of fire.
- Ensure that the Marine's weapons and equipment are prepared for the day's firing as follows:
  - –Supervise the user-serviceability inspection.
  - –Ensure that the Marine's sights are properly blackened.
  - –Verify that the Marines have the proper sight settings on their Service rifles.

–Ensure that the Marines inspect their magazines.

–Ensure that the Marines inspect their ammunition.

–Ensure that the Marines' slings are serviceable, properly assembled, and attached to their Service rifles.

–Ensure that each Marine's gear is properly placed and adjusted.

–Ensure that databooks are properly prepared.

–Ensure that the Marines have hearing protection.

On the Firing Line During Preparation Time

During preparation time on the firing line, the CMC—

- Ensure that the Marines are on assigned firing points.
- Ensure that the Marines check their sight settings.
- Remind the Marines of their target numbers.
- Ensure that the Marines check position and natural point of aim.
- Ensure that Marines dry practice a few shots to practice trigger control.
- Remind the Marines to comply with the four safety rules.

Firing

During firing, the CMC—

- Watches the Marine, not the target.
- Analyzes the Marine's firing positions.
- Reinforces the fundamentals of marksmanship.
- Reinforces compensating for the effects of weather.
- Ensures that the Marines make correct databook entries.
- Analyzes the Marine's performance.
- Ensures that scorekeeping procedures are followed.
- Determines if alibis are warranted and reports them to the block NCO during record firing, when authorized by the range officer.
- Controls movement of the Marines between yard lines and between the firing line and the pits.

After Firing

The following can be performed at the end of a stage of fire or at the end of a day's firing:

- Conduct secondary inspection during unload, show clear.
- Collect excess and/or unused ammunition and turn it in to the appropriate personnel.
- Conduct databook analysis with the Marine.
- Identify Marines for remedial training.
- Issue directions to the Marine for the next stage of fire.

Range Supplies

Range Supplies for a 300-Man Detail.

Nomenclature	NATO Stock Number	Unit of Issue	Quantity
Pasteboard, F prone	6920-00-795-1806	1	Per firing point
Table one			
Target, bull's eye, rifle A	6920-00-627-4071	1	Per firing point
Center, repair target, rifle A	6920-00-627-4072	1	Per Marine
Target, bull's eye, rifle B	6920-00-716-2769	1	Per firing point
Center, target repair, rifle B	6920-00-714-0237	1	Per Marine
Target, rifle D	6920-00-550-7900	1	Per firing point
Center, target repair, rifle D	6920-00-555-9847	1	Per Marine
Target, rifle, B-modified	6920-01-241-5043	1	Per firing point
Center, target repair, rifle B-modified	6920-00-600-6874	1	Per Marine
3-inch spotter	6920-00-713-8255	10	Per firing point
5-inch spotter	6920-00-713-8254	1	Per firing point
10-inch spotter/scoring disk	6920-00-713-8256	2	Per firing point
Spindles for spotters	6920-00-713-8257	13	Per firing point
Black pasties	6920-00-165-6354	1 roll	Per Marine
White pasties	6920-00-172-3572	1 roll	Per Marine
Wheat-paste adhesive	8040-00-275-8105	1 bag	Per 75 targets
Entry-level rifle training data-book	0190-LF-127-7900	1	Per Marine
Annual rifle training data-book	0190-LF-127-0800	1	Per Marine
Scorecard, KD rifle, entry level, line card	0109-LF-064-9100	2	Per Marine
Scorecard, KD rifle, entry level, pit card	0109-LF-064-9200	2	Per Marine
Scorecard, KD rifle, sustainment	0109-LF-066-3900	2	Per Marine

Range Supplies for a 300-Man Detail (Continued).

Nomenclature	NATO Stock Number	Unit of Issue	Quantity
Table two			
Target, rifle, E hardcard	6920-00-795-1806	2	Per Marine
Target, rifle, E hardcard reface	6920-00-554-5054	6	Per Marine
Black pasties	6920-00-165-6354	0.5 roll	Per Marine
3-inch spotter	6920-00-713-8255	6	Per firing point
Spindles for spotters	6920-00-713-8257	6	Per firing point
Score card, table two	—	2	Per Marine
Table three			
Target, rifle, E hardcard	6920-00-795-1806	2	Per Marine
Target, rifle, E hardcard reface	6920-00-554-5054	6	Per Marine
Black pasties	6920-00-165-6354	1 roll	Per Marine
3-inch spotter	6920-00-713-8255	6	Per firing point
Spindles for spotters	6920-00-713-8257	6	Per firing point
Scorecard, table three	—	2	Per Marine
Table four			
Target, rifle, E hardcard	6920-00-795-1806	2	Per Marine
Target, rifle, E hardcard reface	6920-00-554-5054	1	Per Marine
Black pasties	6920-00-165-6354	1 roll	Per Marine
3-inch spotter	6920-00-713-8255	6	Per firing point
Spindles for spotters	6920-00-713-8257	6	Per firing point
Scorecard, table four	—	2	Per Marine

Alibis

An alibi will be awarded during qualification firing if any condition caused by the weapon (i.e., mechanical malfunction), ammunition, or range operation, either line or pit, causes the Marine to not have an equal opportunity to complete a string of fire. An alibi will not be awarded for any condition caused by the Marine.

Circumstances not Constituting an Alibi

The following are examples of events, conditions, or failures to fire that do not constitute an alibi:

- The weapon has not been maintained, cleaned, or lubricated in accordance with Technical Manual 05538/10012-OR, *Operator's Manual with Components List, et. al.*
- The Service rifle or magazine is improperly assembled.
- The failure to—
  - –Seat the magazine properly.
  - –Chamber the first round of each magazine.
  - –Ensure that the bolt is fully forward and locked.
  - –Replace magazines that were determined to be defective during practice.
  - –Shoot the prescribed number of shots for each stage of fire.
  - –Use authorized ammunition.
  - –Adjust the Service rifle sights properly for the string of fire.
  - –Take the weapon off **SAFE** prior to firing.
  - –Engage the safety while firing.
  - –Engage the magazine release button while firing.
  - –Perform remedial action properly when a stoppage occurs. Upon a stoppage, the Marine must execute remedial action.
- The ammunition has been lost.
- The magazines are improperly filled or magazines are not filled with the proper number or rounds.

Circumstances Constituting an Alibi

The following are examples of events, conditions, or failures to fire that constitute an alibi:

- Faulty ammunition.
- A malfunction occurs with the Service rifle that causes a stoppage.
- A fallen or crooked target. If a target is crooked in the frame or falls out after a stage of fire has commenced, the Marine rates an alibi, regardless of whether or not the Marine fires on the target. If the Marine sees that his target is crooked prior to firing, he should inform range personnel.

Firing Tables

Refer to MCO 3574.2_ for the marksmanship firing tables.

Universal BZO Target

A universal BZO target, on page A-11, is used to establish a prezero sight setting at 33 meters/ 36 yards with the Service rifle. The target can be used when employing the RCO or iron sights. Zeroing is discussed in chapter 7.

THIS PAGE INTENTIONALLY LEFT BLANK

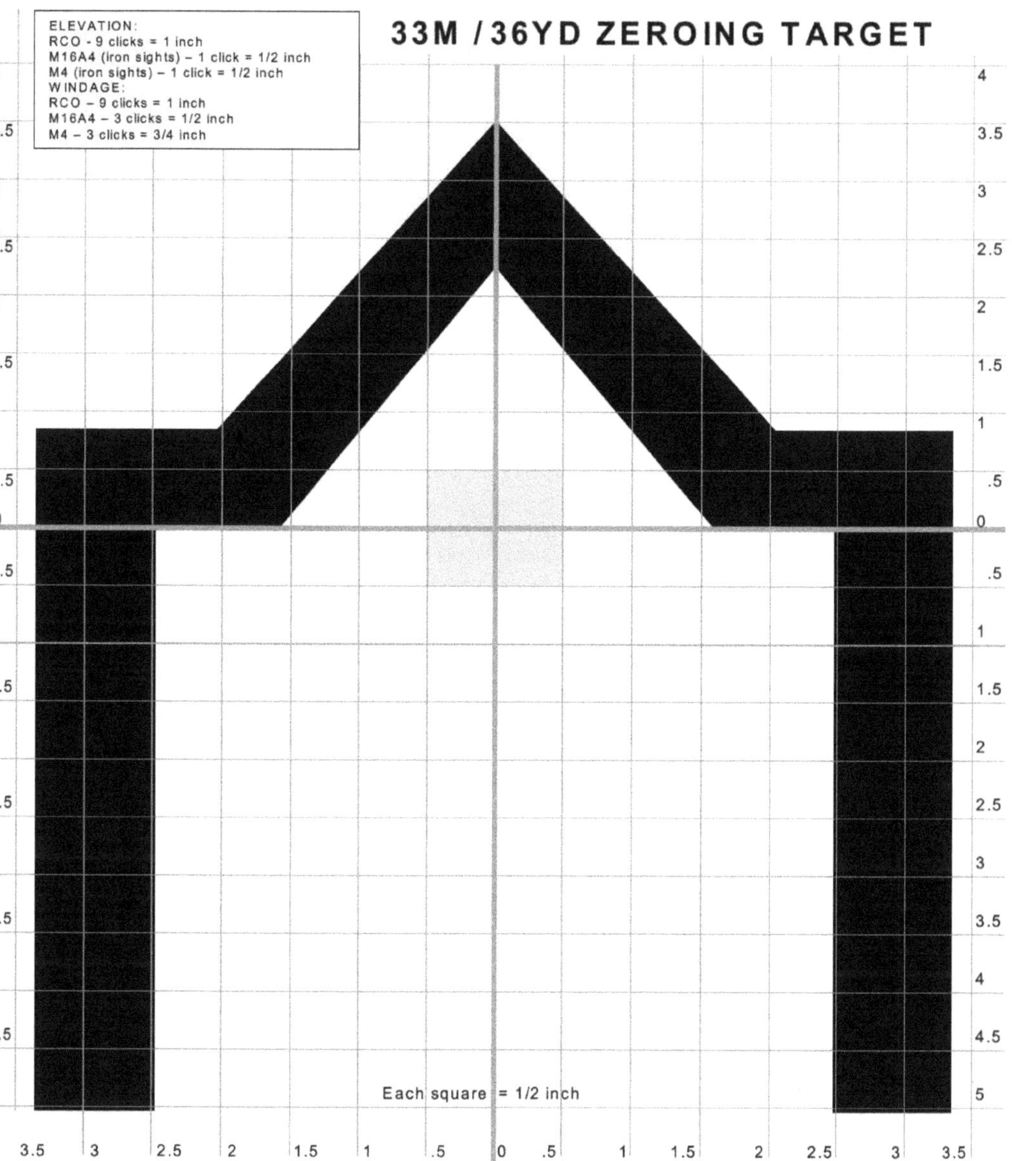
ELEVATION:
RCO - 9 clicks = 1 inch
M16A4 (iron sights) – 1 click = 1/2 inch
M4 (iron sights) – 1 click = 1/2 inch
WINDAGE:
RCO – 9 clicks = 1 inch
M16A4 – 3 clicks = 1/2 inch
M4 – 3 clicks = 3/4 inch
33M / 36YD ZEROING TARGET
4
3.5
3
2.5
2
1.5
1
.5
0
.5
1
1.5
2
2.5
3
3.5
4
4.5
5
Each square = 1/2 inch
3.5 3 2.5 2 1.5 1 .5 0 .5 1 1.5 2 2.5 3 3.5

APPENDIX B Rifle Data Book

The rifle data book is the single most important tool that is available for Marines to evaluate and improve their performance and consistency. The rifle data book is used to record sight adjustments, which enable a zero to be established and maintained. It is critical that all efforts be directed toward establishing a zero setting on the Service rifle that can be taken into combat. The data book provides—

- A precise record of weather conditions and their effect from day to day and a place to record any observations regarding the application of marksmanship fundamentals.
- A documented record of information that Marines analyze to improve shooting performance.
- A ready reference of "must know" information regarding marksmanship.

Types of Data Books

Different data books are created for different types of training. Marines must ensure that the correct data book is used to support the specific training evolution. There are data books for—

- Entry level training.
- Annual rifle training when employing the RCO.

Unlike iron sights, the RCO is not mechanically adjusted to account for changes in elevation and wind. Instead, sight picture is adjusted to support a given firing condition for elevation or wind. The data book pages are designed to reflect this.

Use of the Data Book

As soon as the data book is issued, the following information should be recorded on the data book cover (see fig. B-1 on page B-2):

- *Last name, initials*. Record name for identification purposes.
- *Last 4*. Record the last 4 digits of a social security number for identification purposes.
- *Company*. Record the company to ensure return of a lost data book.
- *Platoon*. Record platoon to ensure return of a lost data book.
- *Blood type*. Record blood type in case an injury occurs on the range.

- *Weapon serial #.* Record the Service rifle serial number to provide a documented record of the Service rifle that was issued from the armory. Each day that a Service rifle is taken from the armory, the serial number of the Service rifle should be compared against the one recorded in the data book.
- *RCO serial #.* Record the RCO's serial number.
- *Range.* Record the firing range.
- *Target.* Record the target that was fired on.
- *Relay.* Record the relay number.
- *Date.* Record the date that the data book was issued.
- *Collimator setting.* Record the collimator setting used as initial sight setting on the RCO.
- *BUIS BZO.* Record the BUIS's BZO setting for elevation and windage.

Before Firing

To save valuable firing time on the range, the BEFORE FIRING section of the data book can be filled out before firing information is recorded and before going to the firing line. See figure B-2. Record the following information in the BEFORE FIRING section of the data book:

- *Zero.* The correct zero for the range fired is shown in the first block of the BEFORE FIRING information. The crotch of the chevron is held center mass on the target at 200 meters/yards.
- *Wind.* Prior to firing, make a wind call by observing your surroundings and by using the flag method and the data book as follows:

ANNUAL RIFLE TRAINING DATABOOK

M16A4 SERVICE RIFLE/M4 CARBINE WITH RIFLE COMBAT OPTIC (RCO) AND BACK-UP IRON SIGHT (BUIS)

LAST NAME, INITIALS:			LAST 4:
UNIT:			BLOOD TYPE:
WEAPON SERIAL #:		RCO SERIAL #:	
RANGE:	TARGET:	RELAY:	DATE:

COLLIMATOR SETTING

ALPHA	NUMERIC

A E I M Q
1 5 9 13 17
3 7 11 15
C G K O

BUIS BZO	ELEV	WIND

NAVMC 11660 (Rev. 11-11) Previous editions are obsolete
S/N: 0109-LF-127-0800 U/I BOX OF 100
FOUO: Privacy sensitive when filled in

Legend:
ELEV elevation

Figure B-1. Data Book Cover Page.

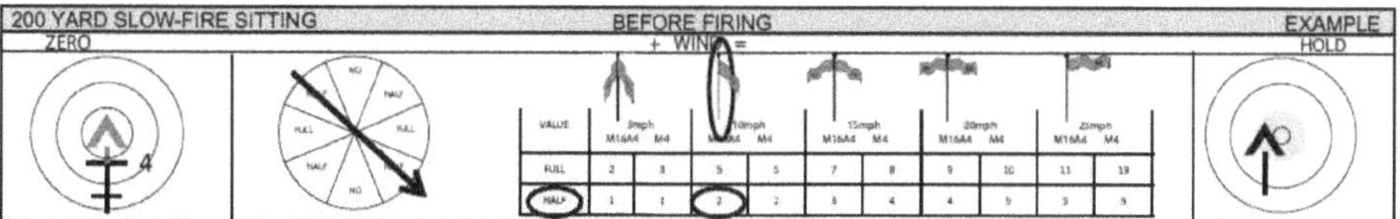

Figure B-2. Data Book: Before Firing Data.

–*Direction*. Determine the direction of the wind and draw an arrow through the clock indicating the direction the wind is blowing in the appropriate sections as follows:

- o Value: under the VALUE column, circle either FULL or HALF to indicate the wind value as indicated by the wind on the clock.
- o Speed: circle the appropriate flag indicating the wind's velocity (i.e., speed).

–*Windage adjustment*. The chart beneath the flag in the different mph columns (i.e., 5, 10, 15, 20, 25) indicates the inches required to offset the effects of the wind. Circle the number where the wind value and wind speed intersect for the weapon being fired.

- o Hold: draw in the hold you will use to compensate for the wind to begin firing.
- o Weather data: see figure B-3.

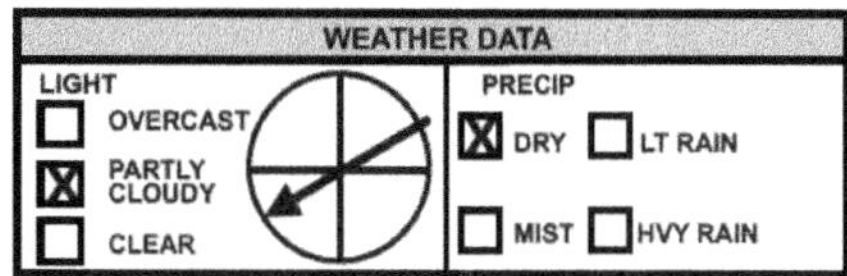

Legend:
HVY heavy
LT light
PRECIP precipitation

Figure B-3. Weather Data.

–*Light*. Under LIGHT check the box that describes the light conditions. Draw an arrow through the clock to indicate the direction the sun is shining. It is important to record light conditions as they affect perception of the target.

–*Precip*. Under PRECIP check the box that describes the precipitation conditions.

During Firing

The DURING FIRING portion of the data book consists of targets used to call and plot shot groups (see fig. B-4 on page B-4). The method for calling and plotting slow fire shots in the data book during slow fire is called the shot behind method. It allows Marines to spend less time recording data and more time firing on the target. This is because all of the calling and plotting is accomplished while the target is in the pits being marked. This information is recorded in the DURING FIRING portion of the data book. The proper and most efficient method for recording data during KD slow fire stages is as follows:

- *Fire the first shot*. Fire the first shot. Then immediately check the wind flag to see if the speed or direction of the wind changed.
- *Call the shot accurately*. As soon as the shot is fired and the target is pulled into the pits, record the exact location of the crotch of the chevron on the target at the exact instant the shot

was fired—record this in the block marked CALL 1. If your hold changed at any time during firing, draw the reticle on the target provided in the block marked HOLD 1. Note that there is an extra (EX) CALL block in case the coach has the Marine fire another shot.

- *Prepare to fire the second shot.* As soon as the call for the first shot has been recorded, prepare to fire the second shot.
- *Look at where the first shot hit.* As the target reappears out of the pits, look where the first shot hit the target. Remember this location so that it can be plotted after firing the second shot.
- *Fire the second shot.* Fire the second shot. Then check the wind flag to see if the wind changed speed or direction.
- *Call the second shot and plot the first shot.* As soon as the second shot is fired and the target is pulled into the pits, record the call of the second shot. Now plot the precise location of the first shot by writing the number 1 on the large target diagram provided in the block marked PLOT.
- *Prepare to fire the third shot.* Repeat these steps until the remaining shots have been fired. Indicate each slow fire shot with the appropriate number (e.g., 1, 2, 3, 4, 5).

After Firing

The bottom portion of the data book is used to record any remarks that will help with the next stage of firing in addition to the sight picture (see fig. B-5). If an adjustment in sight picture is needed, draw the sight picture adjustment on the target in the block marked SIGHT PICTURE ADJUSTMENT (WITHOUT WIND). Record any remarks that will help improve firing in the REMARKS block to include any analysis made after firing by reviewing called and plotted shots.

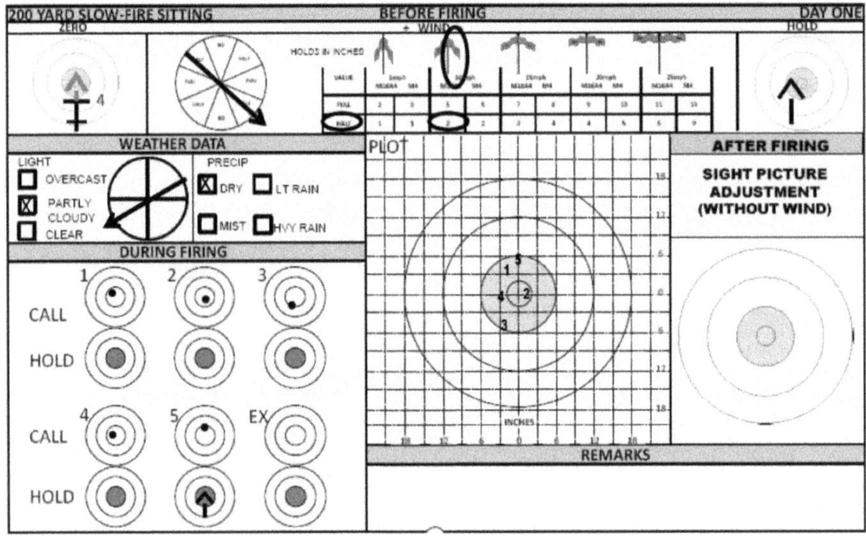

Legend:
EX extra

Figure B-4. Data Book: During Firing Data.

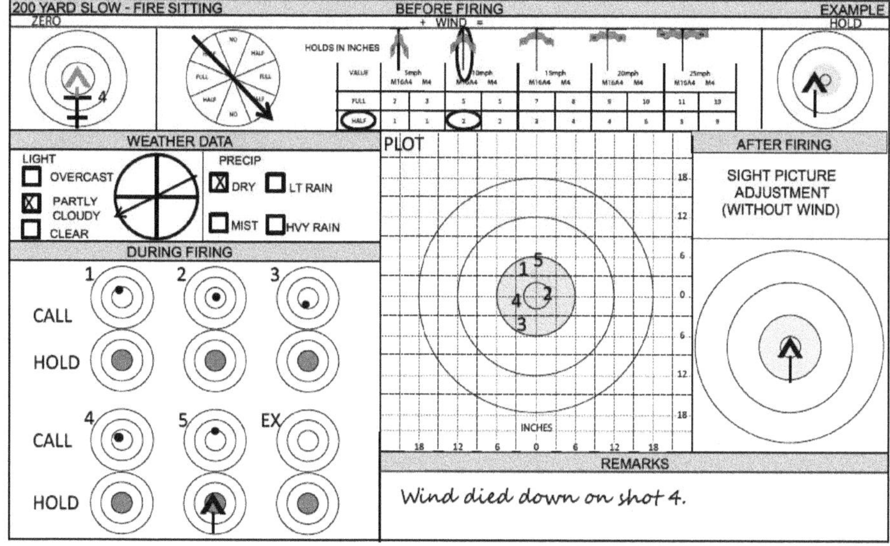

Figure B-5. Data Book: After Firing Data.

Use of the Data Book for Rapid Fire

Before Firing

Record the following information in the BEFORE FIRING section of the data book (see fig. B-6 on page B-6):

- *Zero*. The correct zero for the range fired is shown in the first block of the BEFORE FIRING information. The crotch of the chevron is held center mass on the target at 200 meters/yards.
- *Wind*. Prior to firing, make a wind call by observing your surroundings and by using the flag method and the data book.
  - –*Direction*. Determine the direction of the wind and draw an arrow through the clock, indicating the direction the wind is blowing.
    - o Value: under VALUE, circle FULL or HALF to indicate the wind value as indicated by the wind on the clock.
    - o Speed: circle the appropriate flag indicating the wind's velocity (i.e., speed).
  - –*Windage adjustment*. The chart beneath the flag indicates the inches required to offset the effects of the wind. Circle the number where the wind value and wind speed intersect.
    - o Hold: draw in the hold you will use to compensate for the wind to begin firing.
    - o Weather Data: see figure B-3.

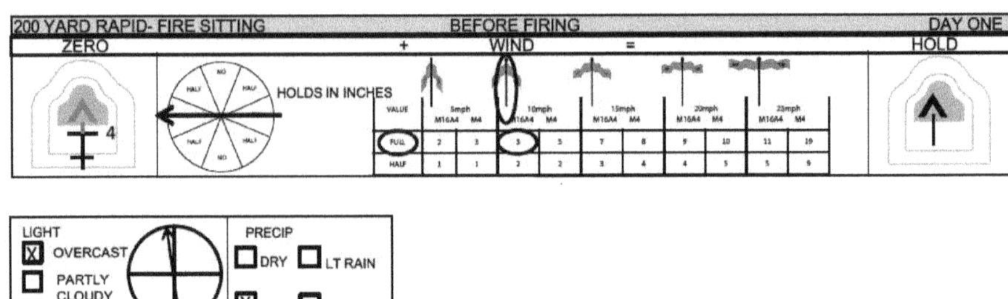

Legend:
HVY heavy
LT light
PRECIP precipitation

Figure B-6. Data Book: Before Firing Data for Rapid Fire.

–*Light*. Under LIGHT, check the box that describes the light conditions. Draw an arrow through the clock to indicate the direction the sun is shining. It is important to record light conditions as they affect perception of the target.

–*Precip*. Under PRECIP, check the box that describes the precipitation conditions.

During Firing

In rapid fire, the firing sequence and the way data are recorded are different from that of slow fire. The following procedures should be used for recording data in the DURING FIRING portion of the data book page for KD rapid fire stages (see fig. B-7):

- *Call shots mentally while firing*. While firing the rapid fire string, call the shots. Make a mental note of any shots called out of the group.
- *Plot all hits with an X*. When the target is marked, plot all visible hits with an X precisely where they appear on the target provided in the block marked PLOT. If a change in hold is required, record it in the block marked 2ND STRING HOLD. Plot all visible hits for the second string with an X on the second target.

After Firing

After firing, record any data or information that may be helpful in improving shooting in the future in the AFTER FIRING section of the data book. See figure B-8. Anything done or observed should be recorded. These items will be helpful when analyzing the daily shooting performance each evening. If an adjustment in sight picture is needed, draw the sight picture adjustment on the target in the block marked SIGHT PICTURE ADJUSTMENT (WITHOUT WIND).

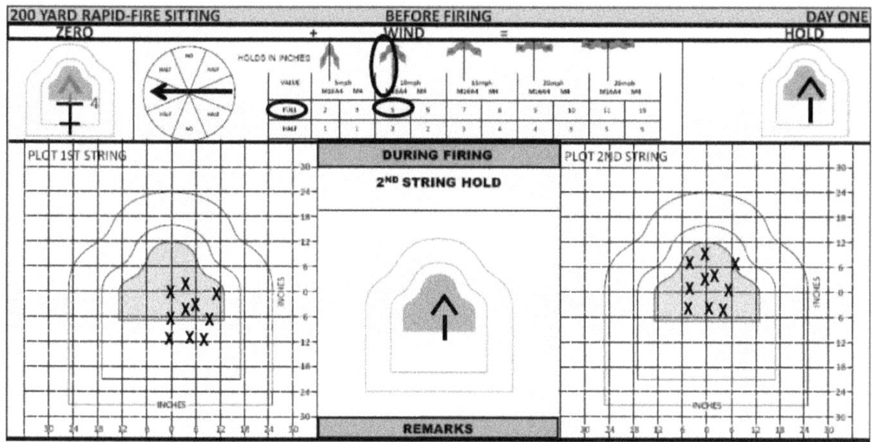

Figure B-7. Data Book: During Firing Data for Rapid Fire.

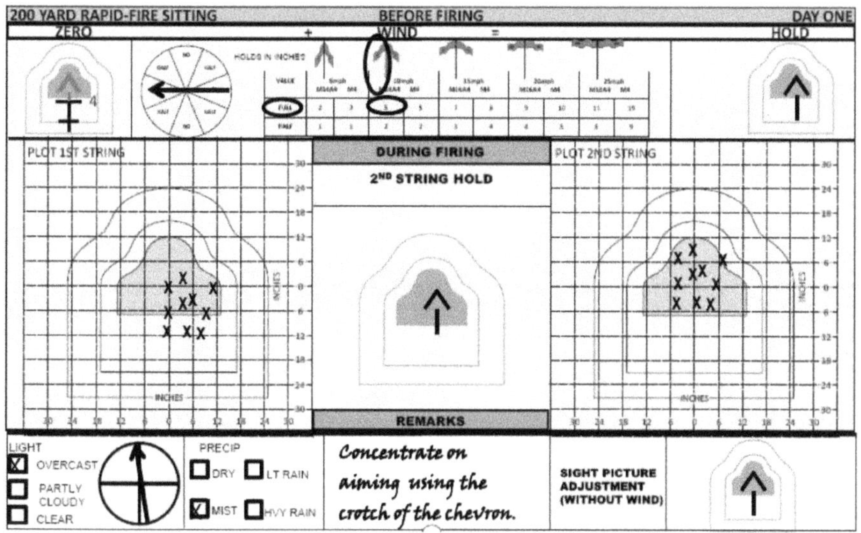

Figure B-8. Data Book: After Firing Data for Rapid Fire.

THIS PAGE INTENTIONALLY LEFT BLANK

Glossary

Section I: Acronyms

BUIS ..... backup iron sight
BZO ..... battlesight zero

CLP ..... cleaner, lubricant, and preservative
CMC ..... combat marksmanship coach

ft ..... foot/feet

IR ..... infrared

KD ..... known distance

m ..... meter(s)
MIL-STD ..... military standard
mm ..... millimeters
mph ..... miles per hour

NCO ..... noncommissioned officer

RCO ..... rifle combat optic

SNCO ..... staff noncommissioned officer

Section II: Definitions

adversary—A party acknowledged as potentially hostile to a friendly party and against which the use of force may be envisaged. (JP 3-0)

aiming point—The precise point where the tip of the front sight post or rifle combat optic reticle pattern is placed in relationship to target.

alert carry—A weapons carry employed with a web sling that is used if enemy contact is likely. The weapon is on SAFE in the Alert carry. The Alert is also used for moving in close terrain.

ambient light—The light produced by natural means (i.e., the sun, moon, and stars).

ambush method—Method used to engage a moving target when it is difficult to track the target with the rifle, as in the prone, sitting, or any supported position.

battlesight zero—Elevation and windage settings required (BUIS) to place a single shot, or the center of a shot group, in a predesignated location on a target from 0 to 300 meters/yards, under ideal weather conditions.

bone support—The body's skeletal structure provides a stable foundation to support the rifle's weight. Bone support is the act of using the body's bones to support as much of the rifle's weight as possible.

box drill—Technique used to engage multiple targets if two shots to the torso fail to eliminate one or both of the threats. In a box drill, the first threat is engaged with a pair to the torso, then the second target is engaged with a pair to the torso. If the threats have not been incapacitated, a shot to the head is fired on the second target, followed by a shot to the head on the first target.

bullet drop compensator—The reticle pattern of the rifle combat optic is a bullet drop compensator with designated aiming points to compensate for trajectory of the 5.56mm round at ranges of 100–800 meters. This feature eliminates the need for mechanical elevation adjustments on the rifle and, instead, uses the reticle pattern for aiming points at each range.

cease fire—Command used to specify when the Marine must stop target engagement.

centerline of the bore—An imaginary straight line beginning at the chamber end of the barrel, proceeding out of the muzzle, and continuing indefinitely.

chamber check—Procedures used to check the condition of a weapon at any time (see Conditions 1, 3, or 4).

clearing barrel—A device that provides a safe direction in which to point a weapon in a controlled environment when loading, unloading, and unloading and showing clear.

Condition 1. Condition of the rifle in which the safety is on, a magazine inserted, a round is in the chamber, the bolt is forward, and the ejection port cover closed.

Condition 3. Condition of the rifle in which the safety is on, a magazine is inserted, the chamber is empty, the bolt is forward, and the ejection port cover closed.

Condition 4. Condition of the rifle in which the safety is on, the magazine is removed, the chamber is empty, the bolt is forward, and the ejection port cover is closed.

condition unknown transfer—The procedures used by the Marine to check the condition of a rifle when he takes charge of the rifle in any situation in which the condition of the rifle is unknown (e.g., an unmanned rifle from a casualty, a rifle stored in a rifle rack).

controlled carry—A weapons carry employed with a three-point sling that is used when no immediate threat is present. The weapon is on **SAFE** in the controlled carry.

controlled pair—Two aimed shots fired upon a target in rapid succession; sight picture is acquired for both shots. A controlled pair is an immediate target engagement technique for targets at greater distances than 15 yards.

corrective action—The process of investigating the cause of a stoppage, clearing the stoppage, and returning the weapon to operation.

eye relief—Distance between the rear sight aperture and the aiming eye.

external ballistics—The study of what a projectile is doing while in flight.

failure to stop—Two shots fired to the torso of a target, followed by an assessment in which it is determined the target has not been incapacitated, followed by a single shot fired to an alternate aiming area.

fire—Command used to specify when a Marine may engage targets.

follow through—The process of maintaining consistency in the support and stability of a weapon during and immediately following delivery of a shot.

function check—A set of procedures performed to ensure the rifle operates properly after the weapon has been reassembled.

hammer pair—Two shots fired in rapid succession with just one sight picture.

hold for rifle combat optics—The placement of the aiming point relative to the target required to place a single shot, or the center of a shot group, in a predesignated location on a target at a specific range, from a specific firing position, under specific weather conditions.

initial sight settings—Rifle elevation and windage settings that serve as the starting point for initial zeroing from which all sight adjustments are made.

internal ballistics—The study of what a projectile is doing while it is inside a weapon.

line of sight—A straight line which begins with the shooter's eye, proceeds through the center of the rear sight aperture, and passes across the tip of the front sight post to a point of aim on a target.

load—Procedures used to take the rifle from Condition 4 to Weapons Condition 3 (see Conditions 1, 2, 3, or 4).

make ready—Procedures used to take the rifle from Condition 3 to Weapons Condition 1 (see Conditions 1, 2, 3, or 4).

natural point of aim—The point at which the rifle sight(s) settle when bone support and muscular relaxation are achieved.

off-center vision—A technique used to search for targets in which the attention is focused on an object without looking directly at it.

ready carry—A weapons carry employed with a web sling that is used if enemy contact is imminent. The weapon is on SAFE in the ready carry.

show clear—Procedures used that require a second party to check the rifle to verify that no ammunition is present before the rifle is placed in Condition 4 (See Conditions 1, 2, 3, or 4).

show clear transfer—A procedure used to transfer a rifle from one Marine to another Marine in which the Marine transferring the rifle leaves the bolt locked to the rear so the Marine receiving the weapon can visually inspect the bore upon receiving it (See Condition 3 and Condition 4).

sight alignment—Relationship between the front sight post and rear sight aperture and the aiming eye.

sight picture—Placement of the tip of the front sight post in relation to the target while maintaining sight alignment.

speed reload—A reload performed when the magazine in the rifle has been emptied and the bolt has locked to the rear. A speed reload is performed as quickly as possible.

stability of hold—The ability to hold the rifle sight(s) still on a designated area of a target.

stock weld—Point of firm contact between the cheek and the stock of the rifle.

tactical reload—A reload performed when the rifle is in Condition 1, before the magazine runs out of ammunition. A tactical reload may be performed when there is a lull in the action.

T-box—An aiming area on the human head, located from the brow to the bottom of the nose and from eye to eye. A shot fired into the T-box is a direct path to the medulla oblongata and is considered an incapacitating shot.

terminal ballistics—The study of what a projectile does once it strikes its target and the effects of a projectile on a target.

three-round burst—When set on BURST, the design of the rifle permits three shots to be fired from a single trigger pull.

tracking method—Method used for engaging a moving target that is moving at a steady pace over a well-determined route.

trajectory—The path a bullet travels in flight to the target.

trigger control—Skillful manipulation of the trigger that causes the rifle to fire without disturbing sight alignment or sight picture.

true zero (backup iron sight)—The elevation and windage settings that are required to place a single shot or the center of a shot group, in a predesignated location on a target at a specific range, from a specific firing position, under ideal weather conditions.

tunnel vision—The singular focus on the threat at the exclusion of peripheral information and overall situational awareness. In tunnel vision, the focus becomes so restrictive or narrow that an indication of other targets is overlooked.

unload—Procedures used to take the rifle from any condition to Condition 4 (see Conditions 1, 2, 3, or 4).

user serviceability inspection—An inspection performed on the rifle before it is fired. This inspection ensures the weapon is in an acceptable operating condition. This inspection does not replace a limited technical inspection or prefire inspection conducted by a qualified armorer.

windage and elevation rules—Define how far the strike of the round will move on the target for each click of front and rear sight elevation or rear sight windage for each 100 meters/yards of range to the target.

zero (rifle combat optic)—Elevation and windage settings required to place a single shot or the center of a shot group in a predesignated location on a target 100 meters/yards, from a specific firing position, under ideal weather conditions.

zero (backup iron sight)—Elevation and windage settings required to place a single shot or the center of a shot group in a predesignated location on a target at a specific range, from a specific firing position, under specific weather conditions.

References and Related Publications

Marine Corps Warfighting Publications (MCWPs)

4-11 Tactical-Level Logistics

5-1 Marine Corps Planning Process

Marine Corps Order (MCO)

3574.2_ Marine Corps Combat Marksmanship Programs

Technical Manuals (TMs)

05538/10012-OR Operator's Manual with Components List for
Rifle, 5.56 MM, M16A2 W/E, NSN: 1005-01-128-9936 (EIC: 4GM) PN 9349000
Rifle, 5.56 MM, M16A4 W/E, NSN: 1005-01-383-2872 (EIC: 4F9) PN 12973001
Carbine, 5.56 MM, M4 W/E, NSN: 1005-01-231-0973 (EIC: 4FJ) PN 9390000
Carbine, 5.56 MM, M4A1 CQBQ W/E, NSN: 1005-01-382-0953 (EIC: 4GC) PN 12972700

10471A-12P/1 Operator's Manual for the Laser Boresight System (LBS)

Military Standard (MIL-STD)

1913 Dimensioning of Accessory Mounting Rail for Small Arms Weapons

Program of Instruction

Hazard Classification of United States Military Explosives and Munitions